Student Solutions Manual

for

Devore and Berk's

Modern Mathematical Statistics
with Applications

Matthew A. Carlton

California Polytechnic State University, San Luis Obispo

BROOKS/COLE
CENGAGE Learning

Australia • Brazil • Japan • Korea • Mexico • Singapore • Spain • United Kingdom • United States

For product information and technology assistance, contact us at **Cengage Learning Customer & Sales Support, 1-800-354-9706**

For permission to use material from this text or product, submit all requests online at **www.cengage.com/permissions** Further permissions questions can be emailed to **permissionrequest@cengage.com**

ISBN-13: 978-0-534-40474-1
ISBN-10: 0-534-40474-X

Brooks/Cole
10 Davis Drive
Belmont, CA 94002-3098
USA

Cengage Learning is a leading provider of customized learning solutions with office locations around the globe, including Singapore, the United Kingdom, Australia, Mexico, Brazil, and Japan. Locate your local office at **www.cengage.com/international**

Cengage Learning products are represented in Canada by Nelson Education, Ltd.

To learn more about Brooks/Cole, visit **www.cengage.com/brookscole**

Purchase any of our products at your local college store or at our preferred online store **www.ichapters.com**

Printed in the United States of America
2 3 4 5 6 7 12 11 10 09 08

CONTENTS

Chapter 1: Overview and Descriptive Statistics

1.

    a.   Houston Chronicle, Des Moines Register, Chicago Tribune, Washington Post

    b.   Capital One, Campbell Soup, Merrill Lynch, Pulitzer

    c.   Bill Jasper, Kay Reinke, Helen Ford, David Menedez

    d.   1.78, 2.44, 3.5, 3.04

3.

    a.   In a sample of 100 VCRs, what are the chances that more than 20 need service while under warranty? What are the chances than none need service while still under warranty?

    b.   What proportion of all VCRs of this brand and model will need service within the warranty period?

5.

    a.   No, the relevant conceptual population is all scores of all students who participate in the SI in conjunction with this particular statistics course.

    b.   The advantage to randomly allocating students to the two groups is that the two groups should then be fairly comparable before the study. If the two groups perform differently in the class, we might attribute this to the treatments (SI and control). If it were left to students to choose, stronger or more dedicated students might gravitate toward SI, confounding the results.

    c.   If all students were put in the treatment group there would be no results with which to compare the treatments.

7.

One could generate a simple random sample of all single family homes in the city or a stratified random sample by taking a simple random sample from each of the 10 district neighborhoods. From each of the homes in the sample the necessary variables would be collected. This would be an enumerative study because there exists a finite, identifiable population of objects from which to sample.

Chapter 1: Overview and Descriptive Statistics

9.

   a. There could be several explanations for the variability of the measurements. Among them could be measuring error, (due to mechanical or technical changes across measurements), recording error, differences in weather conditions at time of measurements, etc.

   b. This could be called a conceptual because there is no sampling frame.

11.

```
6l  |034
6h  |667899
7l  |00122244
7h  |                    Stem=Tens
8l  |001111122344        Leaf=Ones
8h  |5557899
9l  |03
9h  |58
```

This display brings out the gap in the data: There are no scores in the high 70's.

13.

   a.

```
 2 | 23                          stem units: 1.0
 3 | 2344567789                  leaf units: .10
 4 | 01356889
 5 | 00001114455666789
 6 | 000012222334445666677899999
 7 | 00012233455555668
 8 | 02233448
 9 | 012233335666788
10 | 2344455688
11 | 2335999
12 | 37
13 | 8
14 | 36
15 | 0035
16 |
17 |
18 | 9
```

b. A representative value could be the median, 7.0.

c. The data appear to be highly concentrated, except for a few values on the positive side.

d. No, the data is skewed to the right, or positively skewed.

e. The value 18.9 appears to be an outlier, being more than two stem units from the previous value.

15.

a.

Number Nonconforming	Frequency	RelativeFrequency(Freq/60)
0	7	0.117
1	12	0.200
2	13	0.217
3	14	0.233
4	6	0.100
5	3	0.050
6	3	0.050
7	1	0.017
8	1	0.017

doesn't add exactly to 1 because relative frequencies have been rounded 1.001

b. The number of batches with at most 5 nonconforming items is 7+12+13+14+6+3 = 55, which is a proportion of 55/60 = .917. The proportion of batches with (strictly) fewer than 5 nonconforming items is 52/60 = .867. Notice that these proportions could also have been computed by using the relative frequencies: e.g., proportion of batches with 5 or fewer nonconforming items = 1-(.05+.017+.017) = .916; proportion of batches with fewer than 5 nonconforming items = 1 - (.05+.05+.017+.017) = .866.

Chapter 1: Overview and Descriptive Statistics

    c.   The following is a Minitab histogram of this data. The center of the histogram is somewhere around 2 or 3 and it shows that there is some positive skewness in the data. Using the rule of thumb in Exercise 1, the histogram also shows that there is a lot of spread/variation in this data.

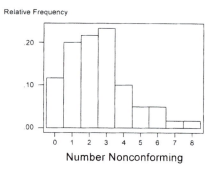

17.

    a.   From this frequency distribution, the proportion of wafers that contained at least one particle is $(100-1)/100 = .99$, or 99%. Note that it is much easier to subtract 1 (which is the number of wafers that contain 0 particles) from 100 than it would be to add all the frequencies for 1, 2, 3,… particles. In a similar fashion, the proportion containing at least 5 particles is $(100 - 1-2-3-12-11)/100 = 71/100 = .71$, or, 71%.

    b.   The proportion containing between 5 and 10 particles is $(15+18+10+12+4+5)/100 = 64/100 = .64$, or 64%. The proportion that contain strictly between 5 and 10 (meaning strictly *more* than 5 and strictly *less* than 10) is $(18+10+12+4)/100 = 44/100 = .44$, or 44%.

    c.   The following histogram was constructed using Minitab. The data was entered using the same technique mentioned in the answer to exercise 8(a). The histogram is *almost* symmetric and unimodal; however, it has a few relative maxima (i.e., modes) and has a very slight positive skew.

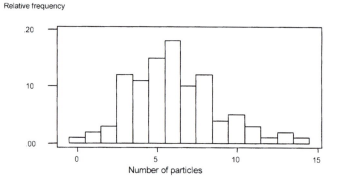

Chapter 1: Overview and Descriptive Statistics

19.

    a. A histogram of the y data appears below. From this histogram, the number of subdivisions having no cul-de-sacs (i.e., y = 0) is 17/47 = .362, or 36.2%. The proportion having at least one cul-de-sac (y ≥ 1) is (47-17)/47 = 30/47 = .638, or 63.8%. Note that subtracting the number of cul-de-sacs with y = 0 from the total, 47, is an easy way to find the number of subdivisions with y ≥ 1.

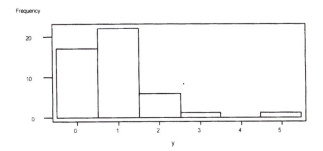

    b. A histogram of the z data appears below. From this histogram, the number of subdivisions with at most 5 intersections (i.e., z ≤ 5) is 42/47 = .894, or 89.4%. The proportion having fewer than 5 intersections (z < 5) is 39/47 = .830, or 83.0%.

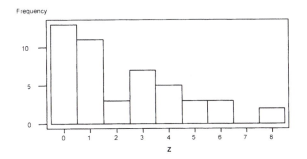

Chapter 1: Overview and Descriptive Statistics

21.

    a.

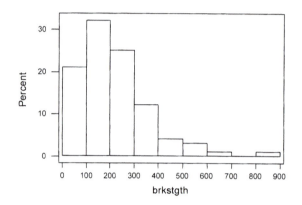

The histogram is skewed right, with a majority of observations between 0 and 300 cycles. The class holding the most observations is between 100 and 200 cycles.

    b.

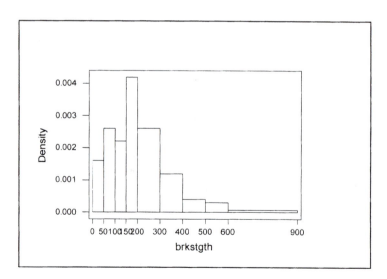

    c.   [proportion $\geq$ 100] = 1 $-$ [proportion < 100] = 1 $-$ .21 = .79

Chapter 1: Overview and Descriptive Statistics

23. Histogram of original data:

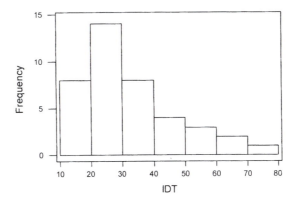

Histogram of transformed data:

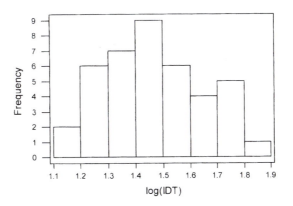

The transformation creates a much more symmetric, mound-shaped histogram.

25.

a.

Class Interval	Frequency	Relative Frequency
0 - 50	9	0.18
50 - 100	19	0.38
100 - 150	11	0.22
150 - 200	4	0.08
200 - 250	2	0.04
250 - 300	2	0.04
300 - 350	1	0.02
350 - 400	1	0.02
400 – 450	0	0.00
450 – 500	0	0.00
500 – 550	1	0.02
	50	1.00

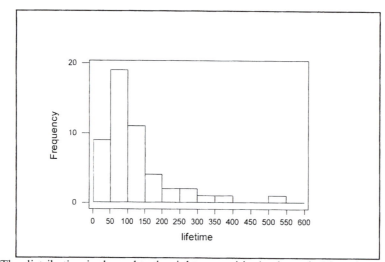

The distribution is skewed to the right, or positively skewed. There is a gap in the histogram, and what appears to be an outlier in the '500 – 550' interval.

b.

Class Interval	Frequency	Relative Frequency
2.25 - 2.75	2	0.04
2.75 - 3.25	2	0.04
3.25 - 3.75	3	0.06
3.75 - 4.25	8	0.16
4.25 - 4.75	18	0.36
4.75 - 5.25	10	0.20
5.25 - 5.75	4	0.08
5.75 - 6.25	3	0.06

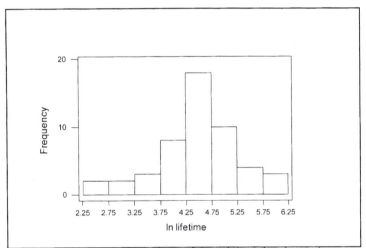

The distribution of the natural logs of the original data is much more symmetric than the original.

c. The proportion of lifetime observations in this sample that are less than 100 is .18 + .38 = .56, and the proportion that is at least 200 is .04 + .04 + .02 + .02 + .02 = .14.

Chapter 1: Overview and Descriptive Statistics

27.

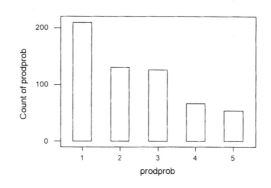

1.   incorrect component
2.   missing component
3.   failed component
4.   insufficient solder
5.   excess solder

29.

a. The frequency distribution is:

Class	Relative Frequency	Class	Relative Frequency
0- 150	.193	900-1050	.019
150- 300	.183	1050-1200	.029
300- 450	.251	1200-1350	.005
450- 600	.148	1350-1500	.004
600- 750	.097	1500-1650	.001
750- 900	.066	1650-1800	.002
		1800-1950	.002

The relative frequency distribution is almost unimodal and exhibits a large positive skew. The typical middle value is somewhere between 400 and 450, although the skewness makes it difficult to pinpoint more exactly than this.

b. The proportion of the fire loads less than 600 is .193+.183+.251+.148 = .775. The proportion of loads that are at least 1200 is .005+.004+.001+.002+.002 = .014.

c. The proportion of loads between 600 and 1200 is 1 - .775 - .014 = .211.

Chapter 1: Overview and Descriptive Statistics

31.

a. The sum of the n = 17 data points is 89, so $\bar{x}$ = 89/17 = 5.24 yards per rush.

b. Since n=17, the median is the (17+1)/2 = 9th position in the list. Sorting, we find the median is 2 yards per rush. The mean differs from the median because the largest observations (rushes of 23 and 24 yards) are much further from the median than are the smallest values.

c. Deleting the smallest (x = −1) and largest (x = 24) values, the sum of the remaining 15 observations is 66. The trimmed mean $\bar{x}_{tr}$ is 66/15 = 4.4 yards per rush. The trimming percentage is 100(1/17) ≈ 5.9%. $\bar{x}_{tr}$ lies between the mean and median.

33.

a. A stem-and leaf display of this data appears below:

```
32|55          stem: ones
33|49          leaf: tenths
34|
35|6699
36|34469
37|03345
38|9
39|2347
40|23
41|
42|4
```

The display is reasonably symmetric, so the mean and median will be close.

b. The sample mean is $\bar{x}$ = 9638/26 = 370.7. The sample median is
$$\tilde{x} = (369+370)/2 = 369.50.$$

c. The largest value (currently 424) could be increased by any amount. Doing so will not change the fact that the middle two observations are 369 and 170, and hence, the median will not change. However, the value x = 424 can not be changed to a number less than 370 (a change of 424-370 = 54) since that *will* lower the values(s) of the two middle observations.

d. Expressed in minutes, the mean is (370.7 sec)/(60 sec) = 6.18 min; the median is 6.16 min.

Chapter 1: Overview and Descriptive Statistics

35.

    a. The reported values are (in increasing order) 110, 115, 120, 120, 125, 130, 130, 135, and 140. Thus the median of the reported values is 125.

    b. 127.6 is reported as 130, so the median is now 130, a very substantial change. When there is rounding or grouping, the median can be highly sensitive to small change.

37.

Using MINITAB, $\tilde{x} = 92$, $\overline{x}_{tr(25)} = 95.38$, $\overline{x}_{tr(10)} = 102.23$, $\overline{x} = 119.3$. All four measures of center have about the same value.

39.

    a. $\overline{y} = \dfrac{\Sigma y_i}{n} = \dfrac{\Sigma(x_i + c)}{n} = \dfrac{\Sigma x_i}{n} + \dfrac{nc}{n} = \overline{x} + c$

       $\tilde{y} = $ the median of $(x_1 + c, x_2 + c,..., x_n + c) = $ median of

       $(x_1, x_2,..., x_n) + c = \tilde{x} + c$

    b. $\overline{y} = \dfrac{\Sigma y_i}{n} = \dfrac{\Sigma(x_i \cdot c)}{n} = \dfrac{c\Sigma x_i}{n} = c\overline{x}$

       $\tilde{y} = $ the median of $(cx_1, cx_2,..., cx_n) = c \cdot median(x_1, x_2,..., x_n) = c\tilde{x}$

41.

    a. range $= 49.3 - 23.5 = 25.8$

    b.

x_i	$(x_i - \overline{x})$	$(x_i - \overline{x})^2$	x_i^2
29.5	-1.53	2.3409	870.25
49.3	18.27	333.7929	2430.49
30.6	-0.43	0.1849	936.36
28.2	-2.83	8.0089	795.24
28.0	-3.03	9.1809	784.00
26.3	-4.73	22.3729	691.69
33.9	2.87	8.2369	1149.21
29.4	-1.63	2.6569	864.36
23.5	-7.53	56.7009	552.25
31.6	0.57	0.3249	998.56

$\Sigma x = 310.3$, $\Sigma(x_i - \overline{x}) = 0$, $\Sigma(x_i - \overline{x})^2 = 443.801$, $\Sigma(x_i^2) = 10{,}072.41$,

$\overline{x} = 31.03$, $s^2 = \dfrac{\sum\limits_{i=1}^{n}(x_i - \overline{x})^2}{n-1} = \dfrac{443.801}{9} = 49.3112$

Chapter 1: Overview and Descriptive Statistics

c. $s = \sqrt{s^2} = 7.0222$

d. $s^2 = \dfrac{\Sigma x^2 - (\Sigma x)^2/n}{n-1} = \dfrac{10,072.41 - (310.3)^2/10}{9} = 49.3112$

43.

a. $\bar{x} = \frac{1}{n}\sum_i x_i = 14438/5 = 2887.6$. The sorted data is: 2781 2856 2888

2900 3013, so the sample median is $\tilde{x} = 2888$.

b. Subtracting a constant from each observation shifts the data, but does not change its sample variance (Exercise 16). For example, by subtracting 2700 from each observation we get the values 81, 200, 313, 156, and 188, which are smaller (fewer digits) and easier to work with. The sum of squares of this transformed data is 204210 and its sum is 938, so the computational formula for the variance gives $s^2 = [204210-(938)^2/5]/(5-1) = 7060.3$.

45.

Using the computational formula, $s^2 = \frac{1}{n-1}\left[\sum_i x_i^2 - \frac{1}{n}\left(\sum_i x_i\right)^2\right] =$

$[3,587,566-(9638)^2/26]/(26-1) = 593.3415$, so s = 24.36. In general, the size of a typical deviation from the sample mean (370.7) is about 24.4. Some observations may deviate from 370.7 by a little more than this, some by less.

47.

First, we need $\bar{x} = \frac{1}{n}\sum x_i = \frac{1}{27}(20,179) = 747.37$. Then we need the sample standard

deviation $s = \sqrt{\dfrac{24,657,511 - \dfrac{(20,179)^2}{27}}{26}} = 606.89$. The maximum award should be

$\bar{x} + 2s = 747.37 + 2(606.89) = 1961.16$, or in dollar units, \$1,961,160. This is quite a bit less than the \$3.5 million that was awarded originally.

49.

Let d denote the fifth deviation. Then $.3+.9+1.0+1.3+d = 0$ or $3.5+d = 0$, so $d = -3.5$. One sample for which these are the deviations is $x_1 = 3.8$, $x_2 = 4.4$, $x_3 = 4.5$, $x_4 = 4.8$, $x_5 = 0$. (obtained by adding 3.5 to each deviation; adding any other number will produce a different sample with the desired property)

Chapter 1: Overview and Descriptive Statistics

51.

   a. The lower half of the data set: 4.4 16.4 22.2 30.0 33.1 36.6, whose median,

and therefore, the lower quartile, is $\dfrac{(22.2 + 30.0)}{2} + 26.1$.

The top half of the data set: 36.6 40.4 66.7 73.7 81.5 109.9, whose median,

and therefore, the upper quartile, is $\dfrac{(66.7 + 73.7)}{2} = 70.2$.

So, the IQR $= (70.2 - 26.1) = 44.1$

   b.

A boxplot (created in Minitab) of this data appears below:

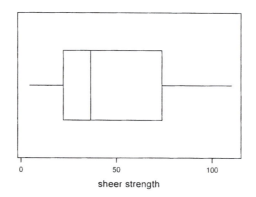

sheer strength

There is a slight positive skew to the data. The variation seems quite large.
There are no outliers.

   c. An observation would need to be further than $1.5(44.1) = 66.15$ units below the
lower quartile $\left[(26.1 - 66.15) = -40.05 \; units\right]$ or above the upper quartile
$\left[(70.2 + 66.15) = 136.35 \; units\right]$ to be classified as a mild outlier. Notice
that, in this case, an outlier on the lower side would not be possible since the
sheer strength variable cannot have a negative value.

An extreme outlier would fall $(3)44.1) = 132.3$ or more units below the lower, or
above the upper quartile. Since the minimum and maximum observations in the
data are 4.4 and 109.9 respectively, we conclude that there are no outliers, of
either type, in this data set.

   d. Not until the value x = 109.9 is lowered below 73.7 would there be any change in
the value of the upper quartile. That is, the value x = 109.9 could not be
decreased by more than $(109.9 - 73.7) = 36.2$ units.

53. A boxplot (created in Minitab) of this data appears below. There is a slight positive skew to this data. There is one extreme outler (x=511). Even when removing the outlier, the variation is still moderately large.

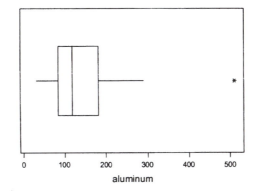

55. The most noticeable feature of the comparative boxplots is that machine 2's sample values have considerably more variation than does machine 1's sample values. However, a typical value, as measured by the median, seems to be about the same for the two machines. The only outlier that exists is from machine 1.

57. A comparative boxplot (created in Minitab) of this data appears below.

The burst strengths for the test nozzle closure welds are quite different from the burst strengths of the production canister nozzle welds.

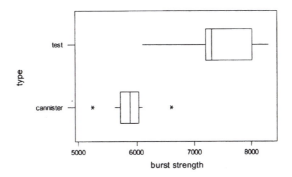

Chapter 1: Overview and Descriptive Statistics

The burst strengths for the test nozzle closure welds are quite different from the burst strengths of the production canister nozzle welds.

The test welds have much higher burst strengths and the burst strengths are much more variable.

The production welds have more consistent burst strength and are consistently lower than the test welds. The production welds data does contain 2 outliers.

59.

From Exercise 39, $\bar{y} = a\bar{x} + b$. So, $S_{yy} = \sum (y_i - \bar{y})^2 = \sum (ax_i + b - [a\bar{x} + b])^2$

$= \sum (ax_i - a\bar{x})^2 = a^2 \sum (x_i - \bar{x})^2 = a^2 S_{xx}$. Thus, $s_y^2 = a^2 s_x^2$ and $s_y = \sqrt{a^2 s_x^2} = |a\, s_x| = |a|s_x$.

61.

Flow rate	Median	Lower quartile	Upper quartile	IQR	1.5(IQR)	3(IQR)
125	3.1	2.7	3.8	1.1	1.65	.3
160	4.4	4.2	4.9	.7	1.05	.1
200	3.8	3.4	4.6	1.2	1.80	3.6

There are no outliers in the three data sets. However, as the comparative boxplot below shows, the three data sets differ with respect to their central values (the medians are different) and the data for flow rate 160 is somewhat less variable than the other data sets. Flow rates 125 and 200 also exhibit a small degree of positive skewness.

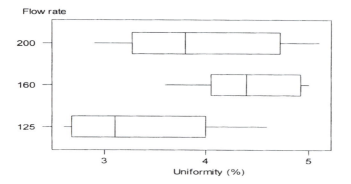

Chapter 1: Overview and Descriptive Statistics

63.

    a.   HC data: $\sum_i x_i^2 = 2618.42$ and $\sum_i x_i = 96.8$,

so $s^2 = [2618.42 - (96.8)^2/4]/3 = 91.953$
and the sample standard deviation is $s = 9.59$.

CO data: $\sum_i x_i^2 = 145645$ and $\sum_i x_i = 735$, so $s^2 = [145645 - (735)^2/4]/3 = 3529.583$ and the sample standard deviation is $s = 59.41$.

    b.   The mean of the HC data is $96.8/4 = 24.2$; the mean of the CO data is $735/4 = 183.75$. Therefore, the coefficient of variation of the HC data is $9.59/24.2 = .3963$, or 39.63%. The coefficient of variation of the CO data is $59.41/183.75 = .3233$, or 32.33%. Thus, even though the CO data has a larger standard deviation than does the HC data, it actually exhibits *less* variability (in percentage terms) around its average than does the HC data.

65.

$$\sum x_i = 163.2$$

$$100\left(\frac{1}{15}\right)\%trimmedmean = \frac{163.2 - 8.5 - 15.6}{13} = 10.70$$

$$100\left(\frac{2}{15}\right)\%trimmedmean = \frac{163.2 - 8.5 - 8.8 - 15.6 - 13.7}{11} = 10.60$$

$$\therefore \frac{1}{2}(100)\left(\frac{1}{15}\right) + \frac{1}{2}(100)\left(\frac{2}{15}\right) = 100\left(\frac{1}{10}\right) = 10\%trimmedmean$$

$$= \frac{1}{2}(10.70) + \frac{1}{2}(10.60) = 10.65$$

67.

    a.   From Exercises 39 and 59, $\bar{y} = a\bar{x} + b$, $s_y^2 = a^2 s_x^2$, and $s_y = |a|s_x$.

    b.   Using the formula $y = 1.8x + 32$, the average temperature was $1.8(38.21)+32 = 100.78°F$ with a standard deviation of $|1.8|(.318) = 0.57°F$.

Chapter 1: Overview and Descriptive Statistics

69.

A table of summary statistics, a stem and leaf display, and a comparative boxplot are below. The healthy individuals have higher receptor binding measure on average than the individuals with PTSD. There is also more variation in the healthy individuals' values. The distribution of values for the healthy is reasonably symmetric, while the distribution for the PTSD individuals is negatively skewed. The box plot indicates that there are no outliers, and confirms the above comments regarding symmetry and skewness.

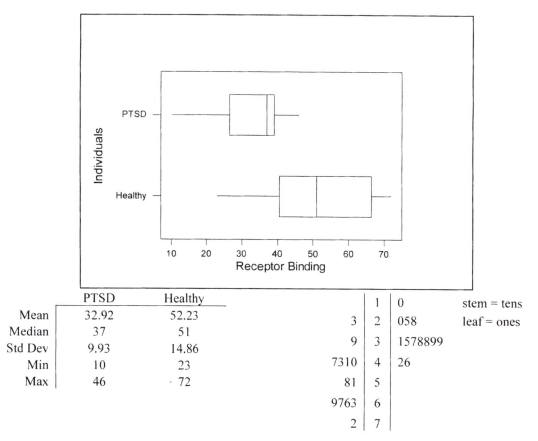

	PTSD	Healthy
Mean	32.92	52.23
Median	37	51
Std Dev	9.93	14.86
Min	10	23
Max	46	· 72

	1	0	stem = tens
3	2	058	leaf = ones
9	3	1578899	
7310	4	26	
81	5		
9763	6		
2	7		

Chapter 1: Overview and Descriptive Statistics

71.

    a.   Mode = .93. It occurs four times in the data set.

    b.   The Modal Category is the one in which the most observations occur.

73.   The measures that are sensitive to outliers are: the mean and the midrange. The mean is sensitive because all values are used in computing it. The midrange is sensitive because it uses only the most extreme values in its computation.

The median, the trimmed mean, and the midfourth are not sensitive to outliers.

The median is the most resistant to outliers because it uses only the middle value (or values) in its computation.

The trimmed mean is somewhat resistant to outliers. The larger the trimming percentage, the more resistant the trimmed mean becomes.

The midfourth, which uses the quartiles, is reasonably resistant to outliers because both quartiles are resistant to outliers.

75.

    a.   Since the constant $\bar{x}$ is subtracted from each x value to obtain each y value, and addition or subtraction of a constant doesn't affect variability, $s_y^2 = s_x^2$ and

$$s_y = s_x$$

    b.   Let $c = 1/s$, where s is the sample standard deviation of the x's and also (by a) of the y's. Then $s_z = c s_y = (1/s)s = 1$, and $s_z^2 = 1$. That is, the "standardized" quantities $z_1, \ldots, z_n$ have a sample variance and standard deviation of 1.

77.

   a.

b.   Proportion less than $20 = \left(\dfrac{216}{391}\right) = .552$

Proportion at least $30 = \left(\dfrac{40}{391}\right) = .102$

c.   First compute $(.90)(391 + 1) = 352.8$. Thus, the 90th percentile should be about the 352nd ordered value. The 351st ordered value lies in the interval 28 - 30. The 352nd ordered value lies in the interval 30 - < 35. There are 27 values in the interval 30 - < 35. We do not know how these values are distributed, however, the smallest value (i.e., the 352nd value in the data set) cannot be smaller than 30. So, the 90th percentile is roughly 30.

d.   First compute $(.50)(391 + 1) = 196$. Thus the median (50th percentile) should be the 196 ordered value. The 174th ordered value lies in the interval 16 - 18. The next 42 observation lie in the interval 18 - < 20. So, ordered observation 175 to 216 lie in the intervals 18 - < 20. The 196th observation is about in the middle of these. Thus, we would say, the median is roughly 19.

79.

    a.   There is some evidence of a cyclical pattern.

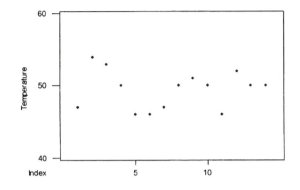

    b.

$$\bar{x}_2 = .1x_2 + .9\bar{x}_1 = (.1)(54) + (.9)(47) = 47.7$$

$$\bar{x}_3 = .1x_3 + .9\bar{x}_2 = (.1)(53) + (.9)(47.7) = 48.23 \approx 48.2, etc.$$

t	$\bar{x}_t\, for\, .\alpha = .1$	$\bar{x}_t\, for\, .\alpha = .5$
1	47.0	47.0
2	47.7	50.5
3	48.2	51.8
4	48.4	50.9
5	48.2	48.4
6	48.0	47.2
7	47.9	47.1
8	48.1	48.6
9	48.4	49.8
10	48.5	49.9
11	48.3	47.9
12	48.6	50.0
13	48.8	50.0
14	48.9	50.0

$\alpha = .1$ gives a smoother series.

c. $\bar{x}_t = \alpha x_t + (1-\alpha)\bar{x}_{t-1}$

$= \alpha x_t + (1-\alpha)[\alpha x_{t-1} + (1-\alpha)\bar{x}_{t-2}]$

$= \alpha x_t + \alpha(1-\alpha)x_{t-1} + (1-\alpha)^2[\alpha x_{t-2} + (1-\alpha)\bar{x}_{t-3}]$

$= \dots = \alpha x_t + \alpha(1-\alpha)x_{t-1} + \alpha(1-\alpha)^2 x_{t-2} + \dots + \alpha(1-\alpha)^{t-2}x_2 + (1-\alpha)^{t-1}\bar{x}_1$

Thus, (x bar)$_t$ depends on x_t and all previous values. As k increases, the coefficient on x_{t-k} decreases (further back in time implies less weight).

d. Not very sensitive, since $(1-\alpha)^{t-1}$ will be very small.

Chapter 2: Probability

1.

  a. Ann is hired but not Bev = A and not B = $A \cap B'$

  b. At least one is hired = A or B (or both) = $A \cup B$

  c. Exactly one is hired = A and not B, or B and not A = $(A \cap B') \cup (B \cap A')$

3.

  a. S = {1324, 1342, 1423, 1432, 2314, 2341, 2413, 2431, 3124, 3142, 4123, 4132, 3214, 3241, 4213, 4231}

  b. Event A contains the outcomes where 1 is first in the list:
   A = {1324, 1342, 1423, 1432}

  c. Event B contains the outcomes where 2 is first or second:
   B = {2314, 2341, 2413, 2431, 3214, 3241, 4213, 4231}

  d. The compound event $A \cup B$ contains the outcomes in A or B or both:
   $A \cup B$ = {1324, 1342, 1423, 1432, 2314, 2341, 2413, 2431, 3214, 3241, 4213, 4231}
   $A \cap B = \varnothing$, since 1 and 2 can't both get into the championship game
   A' = S – A = {2314, 2341, 2413, 2431, 3124, 3142, 4123, 4132, 3214, 3241, 4213, 4231}

5.

  a. Event A = { SSF, SFS, FSS }

  b. Event B = { SSS, SSF, SFS, FSS }

  c. For Event C, the system must have component 1 working (S in the first position), then at least one of the other two components must work (at least one S in the 2nd and 3rd positions: Event C = { SSS, SSF, SFS }

  d. Event C' = { SFF, FSS, FSF, FFS, FFF }
   Event $A \cup C$ = { SSS, SSF, SFS, FSS }
   Event $A \cap C$ = { SSF, SFS }
   Event $B \cup C$ = { SSS, SSF, SFS, FSS }
   Event $B \cap C$ = { SSS SSF, SFS }

Chapter 2: Probability

7.

a.

Outcome Number	Outcome
1	111
2	112
3	113
4	121
5	122
6	123
7	131
8	132
9	133
10	211
11	212
12	213
13	221
14	222
15	223
16	231
17	232
18	233
19	311
20	312
21	313
22	321
23	322
24	323
25	331
26	332
27	333

b. Outcome Numbers 1, 14, 27

c. Outcome Numbers 6, 8, 12, 16, 20, 22

d. Outcome Numbers 1, 3, 7, 9, 19, 21, 25, 27

9.

    a. S = {BBBAAAA, BBABAAA, BBAABAA, BBAAABA, BBAAAAB,
        BABBAAA, BABABAA, BABAABA, BABAAAB, BAABBAA, BAABABA,
        BAABAAB, BAAABBA, BAAABAB, BAAAABB, ABBBAAA, ABBABAA,
        ABBAABA, ABBAAAB, ABABBAA, ABABABA, ABABAAB, ABAABBA,
        ABAABAB, ABAAABB, AABBBAA, AABBABA, AABBAAB, AABABBA,
        AABABAB, AABAABB, AAABBBA, AAABBAB, AAABABB, AAAABBB}

    b. {AAAABBB, AAABABB, AAABBAB, AABAABB, AABABAB}

11.

    a. In the diagram on the left, the shaded area is (A∪B)′. On the right, the shaded
        area is A′, the striped area is B′, and the intersection A′ ∩ B′ occurs where there
        is BOTH shading and stripes. These two diagrams display the same area.

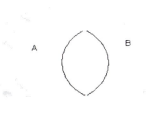
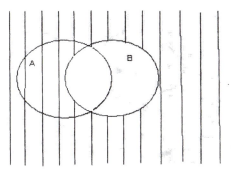

b. In the diagram below, the shaded area represents $(A \cap B)'$. Using the diagram on the right above, the union of A' and B' is represented by the areas that have either shading or stripes or both. Both of the diagrams display the same area.

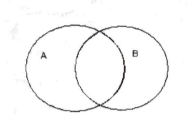

13.

    a. .07

    b. $.15 + .10 + .05 = .30$

    c. Let event A = selected customer owns stocks. Then the probability that a selected customer does not own a stock can be represented by $P(A') = 1 - P(A) = 1 - (.18 + .25) = 1 - .43 = .57$. This could also have been done easily by adding the probabilities of the funds that are not stocks.

15.

    a. awarded either #1 or #2 (or both):
$$P(A_1 \cup A_2) = P(A_1) + P(A_2) - P(A_1 \cap A_2) = .22 + .25 - .11 = .36$$

    b. awarded neither #1 or #2:
$$P(A_1' \cap A_2') = P[(A_1 \cup A_2)'] = 1 - P(A_1 \cup A_2) = 1 - .36 = .64$$

    c. awarded at least one of #1, #2, #3:
$$P(A_1 \cup A_2 \cup A_3) = P(A_1) + P(A_2) + P(A_3) - P(A_1 \cap A_2) - P(A_1 \cap A_3) - $$
$$P(A_2 \cap A_3) + P(A_1 \cap A_2 \cap A_3)$$
$$= .22 + .25 + .28 - .11 - .05 - .07 + .01 = .53$$

    d. awarded none of the three projects:
$$P(A_1' \cap A_2' \cap A_3') = 1 - P(\text{awarded at least one}) = 1 - .53 = .47.$$

e. awarded #3 but neither #1 nor #2:
$$P(\,A_1' \cap A_2' \cap A_3\,) = P(A_3) - P(A_1 \cap A_3) - P(A_2 \cap A_3)$$
$$+ P(A_1 \cap A_2 \cap A_3)$$
$$= .28 - .05 - .07 + .01 \qquad = .17$$

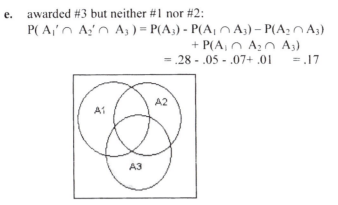

f. either (neither #1 nor #2) or #3:
$$P[(\,A_1' \cap A_2'\,) \cup A_3\,] = P(\text{shaded region}) = P(\text{awarded none}) + P(A_3)$$
$$= .47 + .28 = .75$$

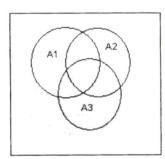

Alternatively, **a – f** can be obtained from the following Venn diagram.

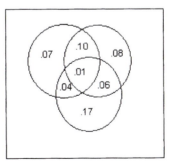

Chapter 2: Probability

17.

    a. Let event E be the event that at most one purchases an electric dryer. Then E′ is the event that at least two purchase electric dryers.

            $P(E′) = 1 - P(E) = 1 - .428 = .572$

    b. Let event A be the event that all five purchase gas. Let event B be the event that all five purchase electric. All other possible outcomes are those in which at least one of each type is purchased. Thus, the desired probability =

            $1 - P(A) - P(B) = 1 - .116 - .005 = .879$

19.

    a. The probabilities do not add to 1 because there are other software packages besides SPSS and SAS for which requests could be made.

    b. $P(A′) = 1 - P(A) = 1 - .30 = .70$

    c. $P(A \cup B) = P(A) + P(B) = .30 + .50 = .80$

            (since A and B are mutually exclusive events)

    d. $P(A′ \cap B′) = P[(A \cup B)′]$ (De Morgan's law)

                $= 1 - P(A \cup B)$

                $= 1 - .80 = .20$

21. Let event A be that the selected joint was found defective by inspector A. $P(A) = \frac{724}{10,000}$. Let event B be analogous for inspector B. $P(B) = \frac{751}{10,000}$. Compound event $A \cup B$ is the event that the selected joint was found defective by at least one of the two inspectors. $P(A \cup B) = \frac{1159}{10,000}$.

    a. The desired event is $(A \cup B)′$, so we use the complement rule:

        $P(A \cup B)′ = 1 - P(A \cup B) = 1 - \frac{1159}{10,000} = \frac{8841}{10,000} = .8841$

    b. The desired event is $B \cap A′$. $P(B \cap A′) = P(B) - P(A \cap B)$.

        $P(A \cap B) = P(A) + P(B) - P(A \cup B)$,

               $= .0724 + .0751 - .1159 = .0316$

        So $P(B \cap A′) = P(B) - P(A \cap B)$

                   $= .0751 - .0316 = .0435$

Chapter 2: Probability

23.

    a. P({M,H}) = .10

    b. P(low auto) = P[{(L,N}, (L,L), (L,M), (L,H)}] = .04 + .06 + .05 + .03 = .18
        Following a similar pattern, P(low homeowner's) = .06 + .10 + .03 = .19

    c. P(same deductible for both) = P[{ LL, MM, HH }] = .06 + .20 + .15 = .41

    d. P(deductibles are different) = 1 – P(same deductibles) = 1 - .41 = .59

    e. P(at least one low deductible) = P[{ LN, LL, LM, LH, ML, HL }]
                               = .04 + .06 + .05 + .03 + .10 + .03 = .31

    f. P(neither low) = 1 – P(at least one low) = 1 - .31 = .69

25. Assume that the computers are numbered 1 – 6 as described. Also assume that computers 1 and 2 are the laptops. Possible outcomes are (1,2) (1,3) (1,4) (1,5) (1,6) (2,3) (2,4) (2,5) (2,6) (3,4) (3,5) (3,6) (4,5) (4,6) and (5,6).

    a. P(both are laptops) = P[{ (1,2)}] = $\frac{1}{15}$ =.067

    b. P(both are desktops) = P[{(3,4) (3,5) (3,6) (4,5) (4,6) (5,6)}] = $\frac{6}{15}$ = .40

    c. P(at least one desktop) = 1 – P(no desktops)
                               = 1 – P(both are laptops)
                               = 1 – .067 = .933

    d. P(at least one of each type) = 1 – P(both are the same)
                               = 1 – P(both laptops) – P(both desktops)
                               = 1 - .067 - .40 = .533

27.

$$P(A \cap B) = P(A) + P(B) - P(A \cup B) = .65$$
$$P(A \cap C) = .55, \ P(B \cap C) = .60$$
$$P(A \cap B \cap C) = P(A \cup B \cup C) - P(A) - P(B) - P(C)$$
$$+ P(A \cap B) + P(A \cap C) + P(B \cap C)$$
$$= .98 - .7 - .8 - .75 + .65 + .55 + .60$$
$$= .53$$

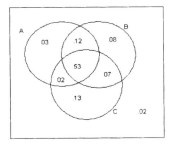

a. $P(A \cup B \cup C) = .98$, as given.

b. P(none selected) $= 1 - P(A \cup B \cup C) = 1 - .98 = .02$

c. P(only automatic transmission selected) $= .03$ from the Venn Diagram

d. P(exactly one of the three) $= .03 + .08 + .13 = .24$

29. There are 27 equally likely outcomes.

a. P(all the same) $= P[(1,1,1)$ or $(2,2,2)$ or $(3,3,3)] = \frac{3}{27} = \frac{1}{9}$

b. P(at most 2 are assigned to the same station) $= 1 - P(\text{all 3 are the same})$
$= 1 - \frac{3}{27} = \frac{24}{27} = \frac{8}{9}$

c. P(all different) $= [\{(1,2,3)\,(1,3,2)\,(2,1,3)\,(2,3,1)\,(3,1,2)\,(3,2,1)\}]$
$= \frac{6}{27} = \frac{2}{9}$

31.

a. $(5)(4) = 20$ (5 choices for president, 4 remain for vice president)

b. $(5)(4)(3) = 60$

c. $\binom{5}{2} = \frac{5!}{2!3!} = 10$ (No ordering is implied in the choice)

33.

a. $(n_1)(n_2) = (9)(27) = 243$

b. $(n_1)(n_2)(n_3) = (9)(27)(15) = 3645$, so such a policy could be carried out for 3645 successive nights, or approximately 10 years, without repeating exactly the same program.

35.

a. $\binom{25}{5} = \frac{25!}{5!20!} = 53,130$

b. $\binom{8}{4} \bullet \binom{17}{1} = 1190$

c. P(exactly 4 have cracks) = $\dfrac{\dbinom{8}{4}\dbinom{17}{1}}{\dbinom{25}{5}} = \dfrac{1190}{53{,}130} = .022$

d. P(at least 4) = P(exactly 4) + P(exactly 5)

$$= \dfrac{\dbinom{8}{4}\dbinom{17}{1}}{\dbinom{25}{5}} + \dfrac{\dbinom{8}{5}\dbinom{17}{0}}{\dbinom{25}{5}} = .022 + .001 = .023$$

37.

There are 10 possible outcomes -- $\dbinom{5}{2}$ ways to select the positions for B's votes:

BBAAA, BABAA, BAABA, BAAAB, ABBAA, ABABA, ABAAB, AABBA, AABAB, and AAABB. Only the last two have A ahead of B throughout the vote count. Since the outcomes are equally likely, the desired probability is $\frac{2}{10} = .20$.

39.

There are $\dbinom{60}{5}$ ways to select the 5 runs. Each catalyst is used in 12 different runs, so the number of ways of selecting one run from each of these 5 groups is 12^5. Thus the desired probability is $\dfrac{12^5}{\dbinom{60}{5}} = .0456$.

41.

a. We want to choose all of the 5 cordless, and 5 of the 10 others, to be among the first 10 serviced, so the desired probability is $\dfrac{\dbinom{5}{5}\dbinom{10}{5}}{\dbinom{15}{10}} = \dfrac{252}{3003} = .0839$

b. Isolating one group, say the cordless phones, we want the other two groups represented in the last 5 serviced. So we choose 5 of the 10 others, except that we don't want to include the outcomes where the last five are all the same.

Chapter 2: Probability

So we have $\dfrac{\binom{10}{5}-2}{\binom{15}{5}}$. But we have three groups of phones, so the desired

probability is $\dfrac{3 \cdot \left[\binom{10}{5}-2\right]}{\binom{15}{5}} = \dfrac{3(250)}{3003} = .2498$.

c. We want to choose 2 of the 5 cordless, 2 of the 5 cellular, and 2 of the corded

phones: $\dfrac{\binom{5}{2}\binom{5}{2}\binom{5}{2}}{\binom{15}{6}} = \dfrac{1000}{5005} = .1998$

43.

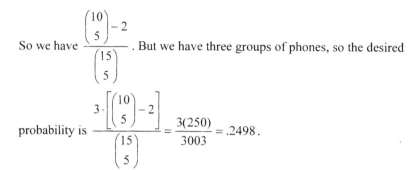

$P(\text{J\&P in 1\&2}) = \dfrac{2 \times 1 \times 4 \times 3 \times 2 \times 1}{6 \times 5 \times 4 \times 3 \times 2 \times 1} = \dfrac{1}{15} = .0667$

$P(\text{J\&P next to each other}) = P(\text{J\&P in 1\&2}) + \ldots + P(\text{J\&P in 5\&6})$
$$= 5 \times \dfrac{1}{15} = \dfrac{1}{3} = .333$$

P(at least one H next to his W) = 1 – P(no H next to his W)
We count the # of ways of no H next to his W as follows:
# of orderings with a H-W pair in seats #1 and 3 and no H next to his W = 6* × 4 × 1*
× 2# × 1 × 1 = 48
*= pair, #=can't put the mate of seat #2 here or else a H-W pair would be in #5 and 6.
# of orderings without a H-W pair in seats #1 and 3, and no H next to his W = 6 × 4 ×
2# × 2 × 1 = 192
#= can't be mate of person in seat #1 or #2.
So, # of seating arrangements with no H next to W = 48 + 192 = 240

And $P(\text{no H next to his W}) = = \dfrac{240}{6 \times 5 \times 4 \times 3 \times 2 \times 1} = \dfrac{1}{3}$, so

$P(\text{at least one H next to his W}) = 1 - \dfrac{1}{3} = \dfrac{2}{3}$

Chapter 2: Probability

45.

a. P(A) = .106 + .141 + .200 = .447, P(C) =.215 + .200 + .065 + .020 = .500 P(A ∩ C) = .200

b. $P(A|C) = \dfrac{P(A \cap C)}{P(C)} = \dfrac{.200}{.500} = .400$. If we know that the individual came from ethnic group 3, the probability that he has type A blood is .40. $P(C|A) = \dfrac{P(A \cap C)}{P(A)} = \dfrac{.200}{.447} = .447$. If a person has type A blood, the probability that he is from ethnic group 3 is .447

c. Define event D = {ethnic group 1 selected}. We are asked for P(D|B′) = $\dfrac{P(D \cap B')}{P(B')} = \dfrac{.192}{.909} = .211$. P(D∩B′)=.082 + .106 + .004 = .192, P(B′) = 1 – P(B) = 1 – [.008 + .018 + .065] = .909

47.

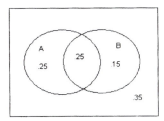

a. $P(B \mid A) = \dfrac{P(A \cap B)}{P(A)} = \dfrac{.25}{.50} = .50$

b. $P(B' \mid A) = \dfrac{P(A \cap B')}{P(A)} = \dfrac{.25}{.50} = .50$

c. $P(A \mid B) = \dfrac{P(A \cap B)}{P(B)} = \dfrac{.25}{.40} = .6125$

d. $P(A' \mid B) = \dfrac{P(A' \cap B)}{P(B)} = \dfrac{.15}{.40} = .3875$

e. $P(A \mid A \cup B) = \dfrac{P[A \cap (A \cup B)]}{P(A \cup B)} = \dfrac{.50}{.65} = .7692$

Chapter 2: Probability

49. The first desired probability is P(both bulbs are 75 watt | at least one is 75 watt).

P(at least one is 75 watt) = 1 – P(none are 75 watt)

$$= 1 - \frac{\binom{9}{2}}{\binom{15}{2}} = 1 - \frac{36}{105} = \frac{69}{105}.$$

Notice that P[(both are 75 watt)∩(at least one is 75 watt)]

$$= P(\text{both are 75 watt}) = \frac{\binom{6}{2}}{\binom{15}{2}} = \frac{15}{105}.$$ So P(both bulbs are 75 watt | at least one is 75

watt) = $\frac{\frac{15}{105}}{\frac{69}{105}} = \frac{15}{69} = .2174$. Second, we want P(same rating | at least one NOT 75

watt). P(at least one NOT 75 watt) = 1 – P(both are 75 watt) = 1 - $\frac{15}{105} = \frac{90}{105}.$

Now, P[(same rating)∩(at least one not 75 watt)] = P(both 40 watt or both 60 watt).

P(both 40 watt or both 60 watt) = $\frac{\binom{4}{2}+\binom{5}{2}}{\binom{15}{2}} = \frac{16}{105}$

Now, the desired conditional probability is $\frac{\frac{16}{105}}{\frac{90}{105}} = \frac{16}{90} = .1778$

51.

   a. P(R from 1st ∩ R from 2nd) = P(R from 2nd | R from 1st) • P(R from 1st)

$$= \frac{8}{11} \bullet \frac{6}{10} = .436$$

   b. P(same numbers)   = P(both selected balls are the same color)

$$= P(\text{both red}) + P(\text{both green}) = .436 + \frac{4}{11} \bullet \frac{4}{10} = .581$$

34

53. $P(B \mid A) = \dfrac{P(A \cap B)}{P(A)} = \dfrac{P(B)}{P(A)}$ (since B is contained in A, $A \cap B = B$) $= \dfrac{.05}{.60} = .0833$

55. $P(A \mid B) + P(A' \mid B) = \dfrac{P(A \cap B)}{P(B)} + \dfrac{P(A' \cap B)}{P(B)} = \dfrac{P(A \cap B) + P(A' \cap B)}{P(B)} = \dfrac{P(B)}{P(B)} = 1$

57. $P(A \cup B \mid C) = \dfrac{P[(A \cup B) \cap C]}{P(C)} = \dfrac{P[(A \cap C) \cup (B \cap C)]}{P(C)}$

$= \dfrac{P(A \cap C) + P(B \cap C) - P(A \cap B \cap C)}{P(C)} = P(A|C) + P(B|C) - P(A \cap B \mid C)$

59.

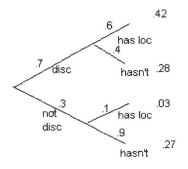

a. P(not disc | has loc) $= \dfrac{P(not.disc \cap has.loc)}{P(has.loc)} = \dfrac{.03}{.03 + .42} = .067$

b. P(disc | no loc) $= \dfrac{P(disc \cap no.loc)}{P(no.loc)} = \dfrac{.28}{.55} = .509$

Chapter 2: Probability

61. Using a tree diagram, B = basic, D = deluxe, W = warranty purchase, W' = no warranty.

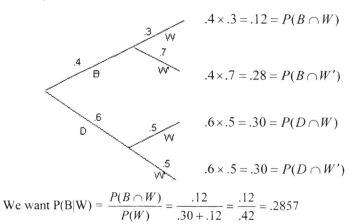

$$.4 \times .3 = .12 = P(B \cap W)$$

$$.4 \times .7 = .28 = P(B \cap W')$$

$$.6 \times .5 = .30 = P(D \cap W)$$

$$.6 \times .5 = .30 = P(D \cap W')$$

We want P(B|W) = $\dfrac{P(B \cap W)}{P(W)} = \dfrac{.12}{.30 + .12} = \dfrac{.12}{.42} = .2857$

63.

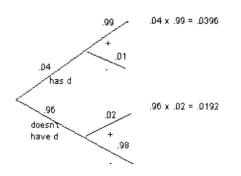

.04 x .99 = .0396

.96 x .02 = .0192

a. P(+) = .0588

b. P(has d | +) = $\dfrac{.0396}{.0588}$ = .6735

c. P(doesn't have d | -) = $\dfrac{.9408}{.9412}$ = .9996

Chapter 2: Probability

65. Define events A1, A2, and A3 as flying with airline 1, 2, and 3, respectively. Events 0, 1, and 2 are 0, 1, and 2 flights are late, respectively. Event DC = the event that the flight to DC is late, and event LA = the event that the flight to LA is late. Creating a tree diagram as described in the hint, the probabilities of the second generation branches are calculated as follows: For the A1 branch, $P(0|A1) = P[DC' \cap LA'] = P[DC'] \cdot P[LA'] = (.7)(.9) = .63$; $P(1|A1) = P[(DC' \cap LA) \cup (DC \cap LA')] = (.7)(.1) + (.3)(.9) = .07 + .27 = .34$; $P(2|A1) = P[DC \cap LA] = P[DC] \cdot P[LA] = (.3)(.1) = .03$. Notice that we have had to assume events DC and LA are *independent* for all airlines. Follow a similar pattern for A2 and A3.

From the law of total probability, we know that

$P(1) = P(A1 \cap 1) + P(A2 \cap 1) + P(A2 \cap 1)$

$= $ (from tree diagram below) $.170 + .105 + .09 = .365$.

We wish to find $P(A1|1)$, $P(A2|1)$, and $P(A2|1)$.

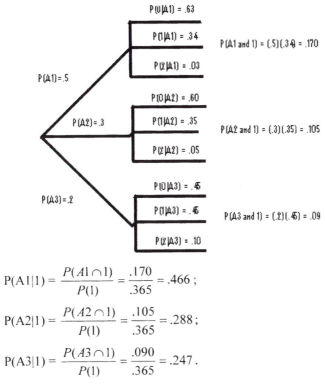

$$P(A1|1) = \frac{P(A1 \cap 1)}{P(1)} = \frac{.170}{.365} = .466 \ ;$$

$$P(A2|1) = \frac{P(A2 \cap 1)}{P(1)} = \frac{.105}{.365} = .288 \ ;$$

$$P(A3|1) = \frac{P(A3 \cap 1)}{P(1)} = \frac{.090}{.365} = .247 \ .$$

Chapter 2: Probability

67.

    a. Since the events are independent, then A′ and B′ are independent, too. (See paragraph below equation 2.7.) $P(B'|A') = . \; P(B') = 1 - .7 = .3$

    b. $P(A \cup B) = P(A) + P(B) - P(A) \cdot P(B) = .4 + .7 - (.4)(.7) = .82$

    c. $P(AB' | A \cup B) = \dfrac{P(AB' \cap (A \cup B))}{P(A \cup B)} = \dfrac{P(AB')}{P(A \cup B)} = \dfrac{.12}{.82} = .146$

69. $P(A' \cap B) = P(B) - P(A \cap B) = P(B) - P(A) \bullet P(B) = [1 - P(A)] \bullet P(B) = P(A') \bullet$
$P(B)$. Alternatively, $P(A' | B) = \dfrac{P(A' \cap B)}{P(B)} = \dfrac{P(B) - P(A \cap B)}{P(B)}$

$= \dfrac{P(B) - P(A) \cdot P(B)}{P(B)} = 1 - P(A) = P(A').$

71. P(no error on any particular question) = .9, so P(no error on any of the 10 questions) $= (.9)^{10} = .3487$. Then P(at least one error) $= 1 - (.9)^{10} = .6513$. For **p** replacing .1, the two probabilities are $(1-\mathbf{p})^n$ and $1 - (1-\mathbf{p})^n$.

73. P(at least one opens) = 1 − P(none open) = $1 - (.05)^5 = .99999969$
P(at least one fails to open) = 1 − P(all open) = $1 - (.95)^5 = .2262$

75.

P(system works) = P(1 − 2 works $\cup$ 3 − 4 works)

              = P(1 − 2 works) + P(3 − 4 works) - P(1 − 2 works $\cap$ 3 − 4 works)

              = P(1 works $\cup$ 2 works) + P(3 works $\cap$ 4 works) − P(1 − 2) $\bullet$ P(3 − 4)

              = (.9+.9-.81) + (.9)(.9) − (.9+.9-.81)(.9)(.9)

              = .99 + .81 - .8019 = .9981

77.

Event A: { (3,1)(3,2)(3,3)(3,4)(3,5)(3,6) }, $P(A) = \frac{1}{6}$;

Event B: { (1,4)(2,4)(3,4)(4,4)(5,4)(6,4) }, $P(B) = \frac{1}{6}$;

Event C: { (1,6)(2,5)(3,4)(4,3)(5,2)(6,1) }, $P(C) = \frac{1}{6}$;

Event A∩B: { (3,4) }; $P(A \cap B) = \frac{1}{36}$;

Chapter 2: Probability

Event $A \cap C$: $\{(3,4)\}$; $P(A \cap C) = \frac{1}{36}$;

Event $B \cap C$: $\{(3,4)\}$; $P(A \cap C) = \frac{1}{36}$;

Event $A \cap B \cap C$: $\{(3,4)\}$; $P(A \cap B \cap C) = \frac{1}{36}$;

$P(A) \cdot P(B) = \frac{1}{6} \cdot \frac{1}{6} = \frac{1}{36} = P(A \cap B)$

$P(A) \cdot P(C) = \frac{1}{6} \cdot \frac{1}{6} = \frac{1}{36} = P(A \cap C)$

$P(B) \cdot P(C) = \frac{1}{6} \cdot \frac{1}{6} = \frac{1}{36} = P(B \cap C)$

The events are pairwise independent.

$P(A) \cdot P(B) \cdot P(C) = \frac{1}{6} \cdot \frac{1}{6} \cdot \frac{1}{6} = \frac{1}{216} \neq \frac{1}{36} = P(A \cap B \cap C)$

The events are not mutually independent

79.

a. Let D_1 = detection on 1st fixation, D_2 = detection on 2nd fixation.

P(detection in at most 2 fixations) = $P(D_1) + P(D_1' \cap D_2)$
$$= P(D_1) + P(D2 \mid D1')P(D_1)$$
$$= p + p(1-p) = p(2-p).$$

b. Define $D_1, D_2, \ldots, D_n$ as in **a**. Then P(at most n fixations)
$$= P(D_1) + P(D_1' \cap D_2) + P(D_1' \cap D_2' \cap D_3) + \ldots + P(D_1' \cap D_2' \cap \ldots \cap D_{n-1}' \cap D_n)$$
$$= p + p(1-p) + p(1-p)^2 + \ldots + p(1-p)^{n-1}$$

$$= p[1 + (1-p) + (1-p)^2 + \ldots + (1-p)^{n-1}] = p \bullet \frac{1-(1-p)^n}{1-(1-p)} = 1 - (1-p)^n$$

Alternatively, P(at most n fixations) = 1 – P(at least n+1 are req'd)
$$= 1 - P(\text{no detection in } 1^{st} \text{ n fixations})$$
$$= 1 - P(D_1' \cap D_2' \cap \ldots \cap D_n')$$
$$= 1 - (1-p)^n$$

c. P(no detection in 3 fixations) = $(1-p)^3$

d. P(passes inspection) = P({not flawed} $\cup$ {flawed and passes})
$$= P(\text{not flawed}) + P(\text{flawed and passes})$$
$$= .9 + P(\text{passes} \mid \text{flawed}) \bullet P(\text{flawed}) = .9 + (1-p)^3(.1)$$

e. P(flawed | passed) = $\dfrac{P(flawed \cap passed)}{P(passed)} = \dfrac{.1(1-p)^3}{.9 + .1(1-p)^3}$

For p = .5, P(flawed | passed) = $\dfrac{.1(.5)^3}{.9 + .1(.5)^3} = .0137$

Chapter 2: Probability

81.

P(system works) = P(1 – 2 works $\cap$ 3 – 4 – 5 – 6 works $\cap$ 7 works)

$\quad\quad$ = P(1 – 2 works) • P(3 – 4 – 5 – 6 works) •P(7 works)

$\quad\quad$ = (.99) (.9639) (.9) = .8588

With the subsystem in figure 2.14 connected in parallel to this subsystem,

P(system works) = .8588+.927 – (.8588)(.927) = .9897

83.

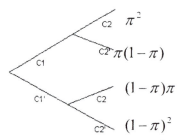

P(at most 1 is lost) = 1 – P(both lost) = 1 – π^2

P(exactly 1 lost) = 2π(1 - π)

P(exactly 1 | at most 1) = $\dfrac{P(exactly 1)}{P(at.most1)} = \dfrac{2\pi(1-\pi)}{1-\pi^2} = \dfrac{2\pi}{1+\pi}$

85.

a. P(line 1) = $\dfrac{500}{1500}$ = .333 ;

$\quad$ P(Crack) = $\dfrac{.50(500)+.44(400)+.40(600)}{1500} = \dfrac{666}{1500}$ = .444

b. P(Blemish | line 1) = .15

c. P(Surface Defect) = $\dfrac{.10(500)+.08(400)+.15(600)}{1500} = \dfrac{172}{1500}$

$\quad$ P(line 1 and Surface Defect) = $\dfrac{.10(500)}{1500} = \dfrac{50}{1500}$

$\quad$ So P(line 1 | Surface Defect) = = $\dfrac{50/1500}{172/1500}$ = .291

Chapter 2: Probability

87.

$P(A \cup B) = P(A) + P(B) - P(A)P(B)$

$.626 \quad = P(A) + P(B) - .144$

So $P(A) + P(B) = .770$ and $P(A)P(B) = .144$.

Let $x = P(A)$ and $y = P(B)$, then using the first equation, $y = .77 - x$, and substituting this into the second equation, we get $x(.77 - x) = .144$ or $x^2 - .77x + .144 = 0$. Use the quadratic formula to solve:

$$\frac{.77 \pm \sqrt{.77^2 - (4)(.144)}}{2} = \frac{.77 \pm .13}{2} = .32 \text{ or } .45$$

So $P(A) = .45$ and $P(B) = .32$

89.

a. There are $5 \times 4 \times 3 \times 2 \times 1 = 120$ possible orderings, so $P(BCDEF) = \frac{1}{120} = .0083$

b. # orderings in which F is $3^{rd} = 4 \times 3 \times 1^* \times 2 \times 1 = 24$, (* because F must be here), so $P(F\ 3^{rd}) = \frac{24}{120} = .2$

c. $P(F \text{ last}) = \dfrac{4 \times 3 \times 2 \times 1 \times 1}{120} = .2$

91. When three experiments are performed, there are 3 different ways in which detection can occur on exactly 2 of the experiments: (i) #1 and #2 and not #3 (ii) #1 and not #2 and #3; (iii) not#1 and #2 and #3. If the impurity is present, the probability of exactly 2 detections in three (independent) experiments is $(.8)(.8)(.2) + (.8)(.2)(.8) + (.2)(.8)(.8) = .384$. If the impurity is absent, the analogous probability is $3(.1)(.1)(.9) = .027$. Thus P(present | detected in exactly 2 out of 3) =

$$\frac{P(\text{detected in exactly } 2 \cap \text{present})}{P(\text{detected in exactly } 2)} = \frac{(.384)(.4)}{(.384)(.4) + (.027)(.6)} = .905$$

93.

a. $P(\text{both} +) = P(\text{carrier} \cap \text{both} +) + P(\text{not a carrier} \cap \text{both} +)$

$= P(\text{both} + | \text{ carrier}) \times P(\text{carrier})$

$\qquad + P(\text{both} + | \text{ not a carrier}) \times P(\text{not a carrier})$

$= (.90)^2(.01) + (.05)^2(.99) = .01058$

$P(\text{both} -) = (.10)^2(.01) + (.95)^2(.99) = .89358$

$P(\text{tests agree}) = .01058 + .89358 = .90416$

b. $P(\text{carrier} | \text{both} + ve) = \dfrac{P(carrier \cap both.positive)}{P(both.positive)} = \dfrac{(.90)^2(.01)}{.01058} = .7656$

95. $P(E_1 \cap \text{late}) = P(\text{ late } | E_1)P(E_1) = (.02)(.40) = .008$

97.   Let B denote the event that a component needs rework.   Then

$$P(B) = \sum_{i=1}^{3} P(B| A_i) \cdot P(A_i) = (.05)(.50) + (.08)(.30) + (.10)(.20) = .069$$

Thus $P(A_1 | B) = \dfrac{(.05)(.50)}{.069} = .362$

$P(A_2 | B) = \dfrac{(.08)(.30)}{.069} = .348$

$P(A_3 | B) = \dfrac{(.10)(.20)}{.069} = .290$

99.

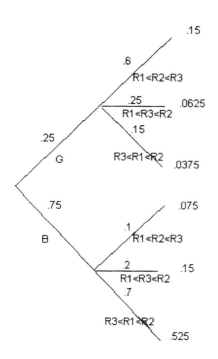

Chapter 2: Probability

a. $P(G \mid R_1 < R_2 < R_3) = \dfrac{.15}{.15 + .075} = .67$, $P(B \mid R_1 < R_2 < R_3) = .33$, classify as granite.

b. $P(G \mid R_1 < R_3 < R_2) = \dfrac{.0625}{.2125} = .2941 < .5$, so classify as basalt.

$P(G \mid R_3 < R_1 < R_2) = \dfrac{.0375}{.5625} = .0667$, so classify as basalt.

c. P(erroneous classif) = P(B classif as G) + P(G classif as B)
= P(classif as G | B)P(B) + P(classif as B | G)P(G)
= $P(R_1 < R_2 < R_3 \mid B)(.75) + P(R_1 < R_3 < R_2 \text{ or } R_3 < R_1 < R_2 \mid G)(.25)$
= (.10)(.75) + (.25 + .15)(.25) = .175

d. For what values of p will $P(G \mid R_1 < R_2 < R_3) > .5$, $P(G \mid R_1 < R_3 < R_2) > .5$, $P(G \mid R_3 < R_1 < R_2) > .5$?

$P(G \mid R_1 < R_2 < R_3) = \dfrac{.6p}{.6p + .1(1 - p)} = \dfrac{.6p}{.1 + .5p} > .5$ iff $p > \dfrac{1}{7}$

$P(G \mid R_1 < R_3 < R_2) = \dfrac{.25p}{.25p + .2(1 - p)} > .5$ iff $p > \dfrac{4}{9}$

$P(G \mid R_3 < R_1 < R_2) = \dfrac{.15p}{.15p + .7(1 - p)} > .5$ iff $p > \dfrac{14}{17}$ (most restrictive)

If $p > \dfrac{14}{17}$ always classify as granite.

101.

a. P(all in correct room) = $\dfrac{1}{4 \times 3 \times 2 \times 1} = \dfrac{1}{24} = .0417$

b. The 9 outcomes which yield incorrect assignments are: 2143, 2341, 2413, 3142, 3412, 3421, 4123, 4321, and 4312, so P(all incorrect) = $\dfrac{9}{24} = .375$

103. Note: s = 0 means that the very first candidate interviewed is hired. Each entry below is the candidate hired for the given policy and outcome.

Outcome	s=0	s=1	s=2	s=3	Outcome	s=0	s=1	s=2	s=3
1234	1	4	4	4	3124	3	1	4	4
1243	1	3	3	3	3142	3	1	4	2
1324	1	4	4	4	3214	3	2	1	4
1342	1	2	2	2	3241	3	2	1	1
1423	1	3	3	3	3412	3	1	1	2
1432	1	2	2	2	3421	3	2	2	1
2134	2	1	4	4	4123	4	1	3	3
2143	2	1	3	3	4132	4	1	2	2
2314	2	1	1	4	4213	4	2	1	3
2341	2	1	1	1	4231	4	2	1	1
2413	2	1	1	3	4312	4	3	1	2
2431	2	1	1	1	4321	4	3	2	1

s	0	1	2	3
P(hire#1)	$\frac{6}{24}$	$\frac{11}{24}$	$\frac{10}{24}$	$\frac{6}{24}$

So s = 1 is best.

105.

$P(A_1) = P(\text{draw slip 1 or 4}) = \frac{1}{2}$; $P(A_2) = P(\text{draw slip 2 or 4}) = \frac{1}{2}$;

$P(A_3) = P(\text{draw slip 3 or 4}) = \frac{1}{2}$; $P(A_1 \cap A_2) = P(\text{draw slip 4}) = \frac{1}{4}$;

$P(A_2 \cap A_3) = P(\text{draw slip 4}) = \frac{1}{4}$; $P(A_1 \cap A_3) = P(\text{draw slip 4}) = \frac{1}{4}$

Hence $P(A_1 \cap A_2) = P(A_1)P(A_2) = \frac{1}{4}$, $P(A_2 \cap A_3) = P(A_2)P(A_3) = \frac{1}{4}$,

$P(A_1 \cap A_3) = P(A_1)P(A_3) = \frac{1}{4}$, thus there exists pairwise independence

$P(A_1 \cap A_2 \cap A_3) = P(\text{draw slip 4}) = \frac{1}{4} \neq 1/8 = P(A_1)p(A_2)P(A_3)$, so the events are not mutually independent.

Chapter 2: Probability

107. The process described in the problem allows for 7 distinct options, which you can picture with an imbalanced tree diagram: $B_0 \cap C' \cap D'$, $B_1 \cap C \cap D$, $B_1 \cap C \cap D'$, $B_1 \cap C' \cap D'$, $B_2 \cap C \cap D$, $B_2 \cap C \cap D'$, $B_2 \cap C' \cap D'$. [Notice, for example, that the combination $C' \cap D$ cannot exist, since a juror won't be dismissed for cause if his bias isn't revealed during voir dire.]

   a. We want $P(B_0|D')$, $P(B_1|D')$, and $P(B_2|D')$. Exhausting all available options listed above, and using the probabilities provided,

$$P(B_0 \mid D') = \frac{P(B_0 \cap D')}{P(D')} = \frac{b_0(1)(1)}{1 - P(D)} = \frac{b_0}{1 - [b_1 cd + b_2 cd]} = \frac{b_0}{1 - (b_1 + b_2)cd}$$

   Similarly, $P(B_i \mid D') = \dfrac{b_i c(1-d) + b_i(1-c)(1)}{P(D')} = \dfrac{b_i(1-cd)}{1 - (b_1 + b_2)cd}$ for $i = 1,2$. It's straightforward to show these sum to 1, using $b_0 + b_1 + b_2 = 1$.

   b. With the numbers provided, $P(B_0|D') = .7117$, $P(B_1|D') = .0577$, $P(B_2|D') = .2306$.

109. Apply the simplest version of Bayes' rule: $P(B|A) > P(B) \rightarrow \dfrac{P(B)P(A|B)}{P(A)} > P(B)$

   $\rightarrow \dfrac{P(A|B)}{P(A)} > 1 \rightarrow P(A|B) > P(A)$.

Chapter 3: Discrete Random Variables and Probability Distributions

1.

S:	FFF	SFF	FSF	FFS	FSS	SFS	SSF	SSS
X:	0	1	1	1	2	2	2	3

3. M = the difference between the large and the smaller outcome with possible values 0, 1, 2, 3, 4, or 5; W = 1 if the sum of the two resulting numbers is even and W = 0 otherwise, a Bernoulli random variable.

5. No. In the experiment in which a coin is tossed repeatedly until a H results, let Y = 1 if the experiment terminates with at most 5 tosses and Y = 0 otherwise. The sample space is infinite, yet Y has only two possible values.

7.

   a. Possible values are 0, 1, 2, ..., 12; discrete

   b. With N = # on the list, values are 0, 1, 2, ... , N; discrete

   c. Possible values are 1, 2, 3, 4, ... ; discrete

   d. $\{ x: 0 < x < \infty \}$ if we assume that a rattlesnake can be arbitrarily short or long; not discrete

   e. With c = amount earned per book sold, possible values are 0, c, 2c, 3c, ... , 10,000c; discrete

   f. $\{ y: 0 < y < 14 \}$ since 0 is the smallest possible pH and 14 is the largest possible pH; not discrete

   g. With m and M denoting the minimum and maximum possible tension, respectively, possible values are $\{ x: m < x < M \}$; not discrete

   h. Possible values are 3, 6, 9, 12, 15, ... -- i.e. 3(1), 3(2), 3(3), 3(4), ...giving a first element, etc,; discrete

9.

   a. Returns to 0 can occur only after an even number of tosses; possible S values are 2, 4, 6, 8, ...(i.e. 2(1), 2(2), 2(3), 2(4),...) an infinite sequence, so X is discrete.

   b. Now a return to 0 is possible after any number of tosses greater than 1, so possible values are 2, 3, 4, 5, ... (1+1,1+2, 1+3, 1+4, ..., an infinite sequence) and X is discrete

11.

a.

x	4	6	8
P(x)	.45	.40	.15

b.

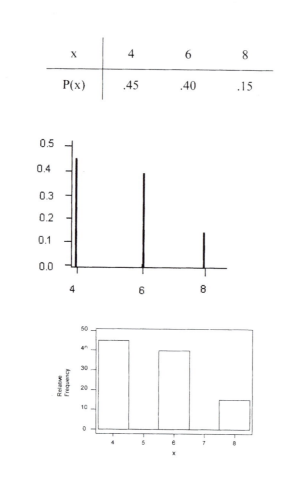

c. $P(x \geq 6) = .40 + .15 = .55$ $\qquad\qquad$ $P(x > 6) = .15$

13.

a. $P(X \leq 3) = p(0) + p(1) + p(2) + p(3) = .10+.15+.20+.25 = .70$
b. $P(X < 3) = P(X \leq 2) = p(0) + p(1) + p(2) = .45$
c. $P(3 \leq X) = p(3) + p(4) + p(5) + p(6) = .55$
d. $P(2 \leq X \leq 5) = p(2) + p(3) + p(4) + p(5) = .71$
e. The number of lines not in use is $6 - X$, so $6 - X = 2$ is equivalent to $X = 4$, $6 - X = 3$ to $X = 3$, and $6 - X = 4$ to $X = 2$. Thus we desire $P(2 \leq X \leq 4) = p(2) + p(3) + p(4) = .65$
f. $6 - X \geq 4$ if $6 - 4 \geq X$, i.e. $2 \geq X$, or $X \leq 2$, and $P(X \leq 2) = .10+.15+.20 = .45$

Chapter 3: Discrete Random Variables and Probability Distributions

15.

  a. (1,2) (1,3) (1,4) (1,5) (2,3) (2,4) (2,5) (3,4) (3,5) (4,5)

  b. $P(X = 0) = p(0) = P[\{ (3,4) (3,5) (4,5)\}] = \frac{3}{10} = .3$

    $P(X = 2) = p(2) = P[\{ (1,2) \}] = \frac{1}{10} = .1$

    $P(X = 1) = p(1) = 1 - [p(0) + p(2)] = .60$, and $p(x) = 0$ if $x \neq 0, 1, 2$

  c. $F(0) = P(X \leq 0) = P(X = 0) = .30$

    $F(1) = P(X \leq 1) = P(X = 0 \text{ or } 1) = .90$

    $F(2) = P(X \leq 2) = 1$

    The c.d.f. is

$$F(x) = \begin{cases} 0 & x < 0 \\ .30 & 0 \leq x < 1 \\ .90 & 1 \leq x < 2 \\ 1 & 2 \leq x \end{cases}$$

17.

  a. $P(2) = P(Y = 2) = P(1^{st} \text{ 2 batteries are acceptable})$
                  $= P(AA) = (.9)(.9) = .81$

  b. $p(3) = P(Y = 3) = P(UAA \text{ or } AUA) = (.1)(.9)^2 + (.1)(.9)^2 = 2[(.1)(.9)^2] = .162$

  c. The fifth battery must be an A, and one of the first four must also be an A. Thus,
    $p(5) = P(AUUUA \text{ or } UAUUA \text{ or } UUAUA \text{ or } UUUAA) = 4[(.1)^3(.9)^2] = .00324$

  d. $P(Y = y) = p(y) = P(\text{the } y^{th} \text{ is an A and so is exactly one of the first } y - 1)$
              $= (y - 1)(.1)^{y-2}(.9)^2, y = 2,3,4,5,\ldots$

19.

Let A denote the type O+ individual (type O positive blood) and B, C, D, the other 3 individuals. Then $p(1) - P(Y = 1) = P(A \text{ first}) = \frac{1}{4} = .25$

$p(2) = P(Y = 2) = P(B, C, \text{ or } D \text{ first and A next}) = \frac{3}{4} \cdot \frac{1}{3} = \frac{1}{4} = .25$

$p(4) = P(Y = 3) = P(A \text{ last}) = \frac{3}{4} \cdot \frac{2}{3} \cdot \frac{1}{2} = \frac{1}{4} = .25$

So $p(3) = 1 - (.25+.25+.25) = .25$

21. The jumps in F(x) occur at x = 0, 1, 2, 3, 4, 5, and 6, so we first calculate F( ) at each of these values:
        $F(0) = P(X \leq 0) = P(X = 0) = .10$
        $F(1) = P(X \leq 1) = p(0) + p(1) = .25$
        $F(2) = P(X \leq 2) = p(0) + p(1) + p(2) = .45$
        $F(3) = .70, F(4) = .90, F(5) = .96, \text{ and } F(6) = 1.$

The c.d.f. is

$$F(x) = \begin{cases} .00 & x < 0 \\ .10 & 0 \le x < 1 \\ .25 & 1 \le x < 2 \\ .45 & 2 \le x < 3 \\ .70 & 3 \le x < 4 \\ .90 & 4 \le x < 5 \\ .96 & 5 \le x < 6 \\ 1.00 & 6 \le x \end{cases}$$

Then $P(X \le 3) = F(3) = .70$, $P(X < 3) = P(X \le 2) = F(2) = .45$,
$P(3 \le X) = 1 - P(X \le 2) = 1 - F(2) = 1 - .45 = .55$,
and $P(2 \le X \le 5) = F(5) - F(1) = .96 - .25 = .71$

23.

  a. Possible X values are those values at which F(x) jumps, and the probability of any particular value is the size of the jump at that value. Thus we have:

x	1	3	4	6	12
p(x)	.30	.10	.05	.15	.40

  b. $P(3 \le X \le 6) = F(6) - F(3-) = .60 - .30 = .30$
  $P(4 \le X) = 1 - P(X < 4) = 1 - F(4-) = 1 - .40 = .60$

25.

  a. Possible X values are 1, 2, 3, …

  $P(1) = P(X = 1) = P(\text{return home after just one visit}) = \frac{1}{3}$

  $P(2) = P(X = 2) = P(\text{second visit and then return home}) = \frac{2}{3} \cdot \frac{1}{3}$

  $P(3) = P(X = 3) = P(\text{three visits and then return home}) = \left(\frac{2}{3}\right)^2 \cdot \frac{1}{3}$

  In general $p(x) = \left(\frac{2}{3}\right)^{x-1}\left(\frac{1}{3}\right)$ for x = 1, 2, 3, …

  b. The number of straight line segments is $Y = 1 + X$ (since the last segment traversed returns Alvie to O), so as in a, $p(y) = \left(\frac{2}{3}\right)^{y-2}\left(\frac{1}{3}\right)$ for y = 2, 3, …

Chapter 3: Discrete Random Variables and Probability Distributions

c. Possible Z values are 0, 1, 2, 3 , …

$p(0) = P(\text{male first and then home}) = \frac{1}{2} \cdot \frac{1}{3} = \frac{1}{6}$,

$p(1) = P(\text{exactly one visit to a female}) = P(\text{female } 1^{st}, \text{ then home}) + P(F, M, \text{home}) + P(M, F, \text{home}) + P(M, F, M, \text{home})$

$= \left(\frac{1}{2}\right)\left(\frac{1}{3}\right) + \left(\frac{1}{2}\right)\left(\frac{2}{3}\right)\left(\frac{1}{3}\right) + \left(\frac{1}{2}\right)\left(\frac{2}{3}\right)\left(\frac{1}{3}\right) + \left(\frac{1}{2}\right)\left(\frac{2}{3}\right)\left(\frac{2}{3}\right)\left(\frac{1}{3}\right)$

$= \left(\frac{1}{2}\right)\left(1 + \frac{2}{3}\right)\left(\frac{1}{3}\right) + \left(\frac{1}{2}\right)\left(\frac{2}{3}\right)\left(\frac{2}{3} + 1\right)\left(\frac{1}{3}\right) = \left(\frac{1}{2}\right)\left(\frac{5}{3}\right)\left(\frac{1}{3}\right) + \left(\frac{1}{2}\right)\left(\frac{2}{3}\right)\left(\frac{5}{3}\right)\left(\frac{1}{3}\right)$

where the first term corresponds to initially visiting a female and the second term corresponds to initially visiting a male. Similarly,

$p(2) = \left(\frac{1}{2}\right)\left(\frac{2}{3}\right)^2 \left(\frac{5}{3}\right)\left(\frac{1}{3}\right) + \left(\frac{1}{2}\right)\left(\frac{2}{3}\right)^2 \left(\frac{5}{3}\right)\left(\frac{1}{3}\right)$. In general,

$p(z) = \left(\frac{1}{2}\right)\left(\frac{2}{3}\right)^{2z-2} \left(\frac{5}{3}\right)\left(\frac{1}{3}\right) + \left(\frac{1}{2}\right)\left(\frac{2}{3}\right)^{2z-2} \left(\frac{5}{3}\right)\left(\frac{1}{3}\right) = \left(\frac{24}{54}\right)\left(\frac{2}{3}\right)^{2z-2}$ for z = 1, 2, 3, …

27. If $x_1 < x_2$, $F(x_2) = P(X \le x_2) = P(\{X \le x_1\} \cup \{x_1 < X \le x_2\}) = P(X \le x_1) + P(x_1 < X \le x_2) \ge P(X \le x_1) = F(x_1)$. $F(x_1) = F(x_2)$ when $P(x_1 < X \le x_2) = 0$.

29.

a. $E(Y) = \sum_{x=0}^{4} y \cdot p(y) = (0)(.60) + (1)(.25) + (2)(.10) + (3)(.05) = .60$

b. $E(100Y^2) = \sum_{x=0}^{4} 100y^2 \cdot p(y) = (0)(.60) + (100)(.25)$

$+ (400)(.10) + (900)(.05) = 110$

31.

a. $E(X) = (13.5)(.2) + (15.9)(.5) + (19.1)(.3) = 16.38$,
$E(X^2) = (13.5)^2(.2) + (15.9)^2(.5) + (19.1)^2(.3) = 272.298$,
$V(X) = 272.298 - (16.38)^2 = 3.9936$

b. $E(25X - 8.5) = 25\,E(X) - 8.5 = (25)(16.38) - 8.5 = 401$

c. $V(25X - 8.5) = V(25X) = (25)^2 V(X) = (625)(3.9936) = 2496$

d. $E[h(X)] = E[X - .01X^2] = E(X) - .01E(X^2) = 16.38 - 2.72 = 13.66$

Chapter 3: Discrete Random Variables and Probability Distributions

33. $E(X) = \sum\limits_{x=1}^{\infty} x \cdot p(x) = \sum\limits_{x=1}^{\infty} x \cdot \dfrac{c}{x^3} = c \sum\limits_{x=1}^{\infty} \dfrac{1}{x^2}$, but it is a well-known result from the

theory of infinite series that $\sum\limits_{x=1}^{\infty} \dfrac{1}{x^2} < \infty$, so $E(X)$ is finite.

35.

P(x)	.8	.1	.08	.02
x	0	1,000	5,000	10,000
H(x)	0	500	4,500	9,500

$E[h(X)] = 600$. Premium should be \$100 plus expected value of damage minus deductible or \$700.

37. $E[h(X)] = E\left(\dfrac{1}{X}\right) = \sum\limits_{x=1}^{6}\left(\dfrac{1}{x}\right) \cdot p(x) = \dfrac{1}{6}\sum\limits_{x=1}^{6}\dfrac{1}{x} = .408$, whereas $\dfrac{1}{3.5} = .286$, so you

expect to win more if you gamble.

39.

    a. The line graph of the p.m.f. of $-X$ is just the line graph of the p.m.f. of X reflected about zero, but both have the same degree of spread about their respective means, suggesting $V(-X) = V(X)$.

    b. With a = -1, b = 0, $V(aX + b) = V(-X) = (-1)^2 V(X) = V(X)$.

41.

    a. $E[X(X-1)] = E(X^2) - E(X)$, $\Rightarrow E(X^2) = E[X(X-1)] + E(X) = 32.5$

    b. $V(X) = 32.5 - (5)^2 = 7.5$

    c. $V(X) = E[X(X-1)] + E(X) - [E(X)]^2$

43.

    a.

k	2	3	4	5	10
$\dfrac{1}{k^2}$	.25	.11	.06	.04	.01

Chapter 3: Discrete Random Variables and Probability Distributions

b. $\mu = \sum\limits_{x=0}^{6} x \cdot p(x) = 2.64$, $\qquad \sigma^2 = \left[\sum\limits_{x=0}^{6} x^2 \cdot p(x) \right] - \mu^2 = 2.37$, $\sigma = 1.54$

Thus $\mu - 2\sigma = -.44$, and $\mu + 2\sigma = 5.72$,
so $P(|x-\mu| \geq 2\sigma) = P(X \text{ is at least 2 s.d.'s from } \mu)$
$\qquad\qquad = P(X \text{ is either} \leq -.44 \text{ or} \geq 5.72) = P(X = 6) = .04$.
Chebyshev's bound of .25 is much too conservative. For k = 3,4,5, and 10, $P(|x-\mu| \geq k\sigma) = 0$, here again pointing to the very conservative nature of the bound $\frac{1}{k^2}$.

c. $\mu = 0$ and $\sigma = \frac{1}{3}$, so $P(|x-\mu| \geq 3\sigma) = P(|X| \geq 1)$

$\qquad\qquad = P(X = -1 \text{ or} +1) = \frac{1}{18} + \frac{1}{18} = \frac{1}{9}$, identical to the upper bound.

d. Let $p(-1) = \frac{1}{50}$, $p(+1) = \frac{1}{50}$, $p(0) = \frac{24}{25}$.

45. $M_X(t) = \Sigma e^{xt} p(x) = \Sigma e^{xt}(.5)^x = \Sigma (.5e^t)^x = \dfrac{.5e^t}{1 - .5e^t}$ or $\dfrac{e^t}{2 - e^t}$, since the sum ranges

from x=1 to x=∞. From this, $E(X) = M'_X(0) = \dfrac{2e^t}{(2 - e^t)^2}\Big|_{t=0} = 2$. Next, $E(X^2) =$

$M''_X(0) = \dfrac{2e^t(2 + e^t)}{(2 - e^t)^3}\Big|_{t=0} = 6$, from which $V(X) = 6 - 2^2 = 2$.

47. Following Example 3.29, $M_X(t) = \Sigma e^{xt} p(x) = \Sigma e^{xt} pq^{x-1} = pe^t \Sigma (e^t q)^{x-1} = pe^t \dfrac{1}{1 - e^t q} =$

$\dfrac{pe^t}{1 - (1 - p)e^t}$. The MGF of Y exactly corresponds to this format with p = .75 (and q = .25). Hence, Y is a geometric random variable with parameter p = .75, and the PMF of Y is $p_Y(y) = .75(.25)^{y-1}$ for y = 1, 2, 3, ...

49. **a.** $M'_X(t) = \exp(5t+2t^2)\cdot(5+4t) \to E(X) = \exp(0)\cdot(5) = 5$. $M''_X(t) = \exp(5t+2t^2)\cdot(5+4t)^2 + \exp(5t+2t^2)\cdot(4) \to E(X^2) = \exp(0)\cdot(5)^2 + \exp(0)\cdot(4) = 25+4 = 29$, from which $V(X) = 29 - 5^2 = 4$.

b. $R_X(t) = 5t+2t^2$. Thus, $R'_X(t) = 5+4t$ and $R''_X(t) = 4$, from which $E(X) = 5+4(0) = 5$ and $V(X) = 4$.

Chapter 3: Discrete Random Variables and Probability Distributions

51. Applying the rescaling proposition, $M_Y(t) = e^{(-5/2)t}M_X((1/2)t) = \exp(-5t/2)\exp(5(t/2)+2(t/2)^2) = \exp(t^2/2)$. Hence, $M_Y'(t) = t\exp(t^2/2)$ and $M_Y''(t) = (t^2+1)\exp(t^2/2)$. From these, $E(Y) = 0$ and $V(Y) = 1 - 0^2 = 1$. (Notice that Y is simply the standardized version of X from Exercise 49.)

53. $E(X) = M_X'(0) = \left.\dfrac{2t}{(1-t^2)^2}\right|_{t=0} = 0$. Next, $E(X^2) = M_X''(0) = \left.\dfrac{6t^2+2}{(1-t^2)^3}\right|_{t=0} = 2$, from which $V(X) = 2 - 0^2 = 2$.

55. All MGFs satisfy $M_X(0) = 1$, but $g(0) = 0$.

57. Work directly with the PMF. $E(X) = 0(.4)+1(.3)+2(.2)+3(.1) = 1$, so $V(X) = (0-1)^2(.4) + (1-1)^2(.3) + (2-1)^2(.2) + (3-1)^2(.1) = 1$ and $\sigma = 1$. Finally, $E[(X - \mu)^3] = (0-1)^3(.4) + (1-1)^3(.3) + (2-1)^3(.2) + (3-1)^3(.1) = .6$, from which $E[(X - \mu)^3]/\sigma^3 = .6$.

59.

   a. $B(4;10,.3) = .850$

   b. $b(4;10,.3) = B(4;10,.3) - B(3;10,.3) = .200$

   c. $b(6;10,.7) = B(6;10,.7) - B(5;10,.7) = .200$

   d. $P(2 \le X \le 4) = B(4;10,.3) - B(1;10,.3) = .701$

   e. $P(2 \le X) = 1 - P(X \le 1) = 1 - B(1;10,.3) = .851$

   f. $P(X \le 1) = B(1;10,.7) = .0000$

   g. $P(2 < X < 6) = P(2 < X \le 5) = B(5;10,.3) - B(2;10,.3) = .570$

61. $X \sim \text{Bin}(6, .10)$

   a. $P(X = 1) = \binom{n}{x}(p)^x(1-p)^{n-x} = \binom{6}{1}(.1)^1(.9)^5 = .3543$

   b. $P(X \ge 2) = 1 - [P(X = 0) + P(X = 1)]$.

   From **a**, we know $P(X = 1) = .3543$, and $P(X = 0) = \binom{6}{0}(.1)^0(.9)^6 = .5314$.

   Hence $P(X \ge 2) = 1 - [.3543 + .5314] = .1143$

c. Either 4 or 5 goblets must be selected

i) Select 4 goblets with zero defects: $P(X = 0) = \binom{4}{0}(.1)^0(.9)^4 = .6561$.

ii) Select 4 goblets, one of which has a defect, and the 5th is good:

$$\left[\binom{4}{1}(.1)^1(.9)^3\right] \times .9 = .26244$$

So the desired probability is $.6561 + .26244 = .91854$

63. Let S = has at least one citation. Then p = .4, n = 15
   a. If at least 10 have no citations (Failure), then at most 5 have had at least one (Success):
   $$P(X \le 5) = B(5;15,.40) = .403$$

   b. $P(X \le 7) = B(7;15,.40) = .787$

   c. $P(5 \le X \le 10) = P(X \le 10) - P(X \le 4) = .991 - .217 = .774$

65. Let S represent a telephone that is submitted for service while under warranty and must be replaced. Then $p = P(S) = P(\text{replaced} \mid \text{submitted}) \cdot P(\text{submitted}) = (.40)(.20) = .08$. Thus X, the number among the company's 10 phones that must be replaced, has a binomial distribution with n = 10, p = .08, so $p(2) = P(X=2) =$

$$\binom{10}{2}(.08)^2(.92)^8 = .1478$$

67. X = the number of flashlights that work. Let event B = {battery has acceptable voltage}. Then $P(\text{flashlight works}) = P(\text{both batteries work}) = P(B)P(B) = (.9)(.9) = .81$ We must assume that the batteries' voltage levels are independent. $X \sim \text{Bin}(10, .81)$. $P(X \ge 9) = P(X=9) + P(X=10) =$

$$\binom{10}{9}(.81)^9(.19) + \binom{10}{10}(.81)^{10} = .285 + .122 = .407$$

69.

   a. $P(\text{rejecting claim when } p = .8) = B(15;25,.8) = .017$

   b. $P(\text{not rejecting claim when } p = .7) = P(X \ge 16 \text{ when } p = .7)$
   $= 1 - B(15;25,.7) = 1 - .189 = .811$; for p = .6, this probability is
   $= 1 - B(15;25,.6) = 1 - .575 = .425$.

   c. The probability of rejecting the claim when p = .8 becomes $B(14;25,.8) = .006$, smaller than in **a** above. However, the probabilities of **b** above increase to .902 and .586, respectively.

Chapter 3: Discrete Random Variables and Probability Distributions

71. If topic A is chosen, when n = 2, P(at least half received)
= P(X ≥ 1) = 1 – P(X = 0) = 1 – $(.1)^2$ = .99
If B is chosen, when n = 4, P(at least half received)
= P(X ≥ 2) = 1 – P(X ≤ 1) = 1 – $(0.1)^4$ – $4(.1)^3(.9)$ = .9963
Thus topic B should be chosen.
If p = .5, the probabilities are .75 for A and .6875 for B, so now A should be chosen.

73.

   a. b(x; n, 1 – p) = $\binom{n}{x}(1-p)^x(p)^{n-x}$ = $\binom{n}{n-x}(p)^{n-x}(1-p)^x$ = b(n-x; n, p)

       Alternatively, P(x S's when P(S) = 1 – p) = P(n-x F's when P(F) = p), since the
two events are identical, but the labels S and F are arbitrary and so can be
interchanged (if P(S) and P(F) are also interchanged), yielding P(n-x S's when
P(S) = 1 – p) as desired.

   b. B(x;n,1 – p) = P(at most x S's when P(S) = 1 – p)
                 = P(at least n-x F's when P(F) = p)
                 = P(at least n-x S's when P(S) = p)
                 = 1 – P(at most n-x-1 S's when P(S) = p)
                 = 1 – B(n-x-1;n,p)

   c. Whenever p > .5, (1 – p) < .5 so probabilities involving X can be calculated using
the results **a** and **b** in combination with tables giving probabilities only for p ≤ .5

75.

   a. Although there are three payment methods, we are only concerned with S = uses
a debit card and F = does not use a debit card. Thus we can use the binomial
distribution. So n = 100 and p = .2. E(X) = np = 100(.5) = 20, and V(X) = 16.

   b. With S = doesn't pay with cash, n = 100 and p = .7, E(X) = np = 100(.7) = 70,
and V(X) = 21.

77.

     When p = .5, μ = 10 and σ = 2.236, so 2σ = 4.472 and 3σ = 6.708.
The inequality |X – 10| ≥ 4.472 is satisfied if either X ≤ 5 or X ≥ 15, or P(|X - μ| ≥ 2σ)
= P(X ≤ 5 or X ≥ 15) = .021 + .021 = .042. The inequality |X – 10| ≥ 6.708 is satisfied
if either X ≤ 3 or X ≥ 17, so P(|X – μ| ≥ 3σ) = P(X ≤3 or X ≥ 17) = .001 + .001 = .002.
In the case p = .75, μ = 15 and σ = 1.937, so 2σ = 3.874 and 3σ = 5.811. P(|X - 15| ≥
3.874) = P(X ≤ 11 or X ≥ 19) = .041 + .024 = .065, whereas P(|X - 15| ≥ 5.811) =
P(X ≤ 9) = .004.
All these probabilities are considerably less than the upper bounds .25 (for k = 2) and
.11 (for k = 3) given by Chebyshev.

Chapter 3: Discrete Random Variables and Probability Distributions

79. Let $Y = n - X = -1X + n$. Then $M_Y(t) = e^{nt} M_X(-t) = e^{nt}(pe^{-t} + q)^n = [e^t(pe^{-t} + q)]^n = (qe^t + p)^n$. We can find the mean and variance of Y directly, or recognize this as the MGF of a Binomial random variable with <u>success</u> parameter q ($= 1 - p$). Hence, $E(Y) = nq$ and $V(Y) = nqp$.

These answers are intuitive for two reasons. Logically, we can interchange the roles of "success" and "failure" to make Y a binomial random variable, with the feature that $E(Y) = nq = n(1 - p) = n - np = n - E(X)$. Also, since Y is a rescaling of X with scale coefficient -1, we know we should have $V(Y) = V(X)$, and we do.

81. $X \sim h(x; 6, 12, 7)$

   a. $P(X=5) = \dfrac{\binom{7}{5}\binom{5}{1}}{\binom{12}{6}} = \dfrac{105}{924} = .114$

   b. $P(X \leq 4) = 1 - P(X \geq 5) = 1 - [P(X=5) + P(X=6)] =$

   $1 - \left[\dfrac{\binom{7}{5}\binom{5}{1}}{\binom{12}{6}} + \dfrac{\binom{7}{6}}{\binom{12}{6}}\right] = 1 - \dfrac{105 + 7}{924} = 1 - .121 = .879$

   c. $E(X) = \left(\dfrac{6 \cdot 7}{12}\right) = 3.5$; $\sigma = \sqrt{\left(\frac{6}{11}\right)(6)\left(\frac{7}{12}\right)\left(\frac{5}{12}\right)} = \sqrt{.795} = .892$

   $P(X > 3.5 + .892) = P(X > 4.392) = P(X \geq 5) = .121$ (see part b)

   d. We can approximate the hypergeometric distribution with the binomial if the population size and the number of successes are large: $h(x;15,40,400)$ approaches $b(x;15,.10)$. So $P(X \leq 5) \approx B(5; 15, .10)$ from the binomial tables $= .998$

83.

   a. Possible values of X are 5, 6, 7, 8, 9, 10. (In order to have less than 5 of the granite, there would have to be more than 10 of the basaltic).

   $P(X = 5) = h(5; 15,10,20) = \dfrac{\binom{10}{5}\binom{10}{10}}{\binom{20}{15}} = .0163$.

   Following the same pattern for the other values, we arrive at the pmf, in table form below.

x	5	6	7	8	9	10
p(x)	.0163	.1354	.3483	.3483	.1354	.0163

b. P(all 10 of one kind or the other) = P(X = 5) + P(X = 10) = .0163 + .0163 = .0326

c. $E(X) = n \cdot \dfrac{M}{N} = 15 \cdot \dfrac{10}{20} = 7.5$; $V(X) = \left(\dfrac{5}{19}\right)(7.5)\left(1 - \dfrac{10}{20}\right) = .9868$;

$\sigma_x = .9934$

$\mu \pm \sigma = 7.5 \pm .9934 = (6.5066, 8.4934)$, so we want
P(X = 7) + P(X = 8) = .3483 + .3483 = .6966

85.

a. h(x; 10,10,20) (the successes here are the top 10 pairs, and a sample of 10 pairs is drawn from among the 20)

b. Let X = the number among the top 5 who play E-W. Then P(all of top 5 play the same direction) = P(X = 5) + P(X = 0) = h(5;10,5,20) + h(5;10,5,20)

$$= \frac{\dbinom{15}{5}}{\dbinom{20}{10}} + \frac{\dbinom{15}{10}}{\dbinom{20}{10}} = .033$$

c. N = 2n; M = n; n = n
h(x;n,n,2n)

$E(X) = n \cdot \dfrac{n}{2n} = \dfrac{1}{2} n$;

$V(X) = \left(\dfrac{2n - n}{2n - 1}\right) \cdot n \cdot \dfrac{n}{2n} \cdot \left(1 - \dfrac{n}{2n}\right) = \left(\dfrac{n}{2n - 1}\right) \cdot \dfrac{n}{2} \cdot \left(1 - \dfrac{n}{2n}\right) = \left(\dfrac{n}{2n - 1}\right) \cdot \dfrac{n}{2} \cdot \left(\dfrac{1}{2}\right)$

87.

a. With S = a female child and F = a male child, let X = the number of F's before the 2nd S. Then P(X = x) = nb(x;2, .5)

b. P(exactly 4 children) = P(exactly 2 males)
= nb(2;2,.5) = (3)(.0625) = .188

c. P(at most 4 children) = P(X ≤ 2)

$$= \sum_{x=0}^{2} nb(x;2,.5) = .25 + 2(.25)(.5) + 3(.0625) = .688$$

d. $E(X) = \dfrac{(2)(.5)}{.5} = 2$, so the expected number of children = E(X + 2) = 4

Chapter 3: Discrete Random Variables and Probability Distributions

89. This is identical to an experiment in which a single family has children until exactly 6 females have been born(since p = .5 for each of the three families), so p(x) = nb(x;6,.5) and E(X) = 6 (= 2+2+2, the sum of the expected number of males born to each one.)

91. $M_X(t) = \dfrac{p^r}{(1-qe^t)^r}$, so $M'_X(0) = \dfrac{rqp^r e^t}{(1-qe^t)^{r+1}}\bigg|_{t=0} = \dfrac{rqp^r}{(1-q)^{r+1}} = \dfrac{rqp^r}{p^{r+1}} = \dfrac{rq}{p}$. Next,

$E(X^2) = M''_X(0) = \dfrac{rqp^r[rqe^{2t}+e^t]}{(1-qe^t)^{r+2}}\bigg|_{t=0} = \cdots = \dfrac{rq(rq+1)}{p^2}$, from which $V(X) =$

$\dfrac{rq(rq+1)}{p^2} - \left[\dfrac{rq}{p}\right]^2 = \dfrac{rq}{p^2}$.

93.

    a. P(X ≤ 8) = F(8;5) = .932

    b. P(X = 8) = F(8;5) - F(7;5) = .065

    c. P(X ≥ 9) = 1 – P(X ≤ 8) = .068

    d. P(5 ≤ X ≤ 8) = F(8;5) – F(4;5) = .492

    e. P(5 < X < 8) = F(7;5) – F(5;5) = .867-.616=.251

95.

    a. P(X ≤ 5) = F(5;8) = .191

    b. P(6 ≤ X ≤ 9) = F(9;8) – F(5;8) = .526

    c. P(X ≥ 10) = 1 - P(X ≤ 9) = .283

    d. E(X) = λ = 8, $\sigma_X = \sqrt{\lambda} = 2.83$, so P(X > 10.83) = P(X ≥ 11) = 1 – P(X ≤ 10) = 1 - .816 = .184

97. $p = \dfrac{1}{200}$; n = 1000; λ = np = 5

    a. P(5 ≤ X ≤ 8) = F(8;5) – F(4;5) = .492

    b. P(X ≥ 8) = 1 – P(X ≤ 7) = 1 - .867 = .133

Chapter 3: Discrete Random Variables and Probability Distributions

99.

    a. $\lambda = 8$ when $t = 1$, so $P(X = 6) = F(6;8) - F(5;8) = .313 - .191 = .122$,
              $P(X \geq 6) = 1 - F(5;8) = .809$, and $P(X \geq 10) = 1 - F(9;8) = .283$

    b. $t = 90$ min $= 1.5$ hours, so $\lambda = 12$; thus the expected number of arrivals is 12 and
        the SD $= \sqrt{12} = 3.464$

    c. $t = 2.5$ hours implies that $\lambda = 20$; in this case, $P(X \geq 20) = 1 - F(19;20) = .530$
        and $P(X \leq 10) = F(10;20) = .011$.

101.

    a. For a two hour period the parameter of the distribution is $\lambda = \alpha t = (4)(2) = 8$,
        so $P(X = 10) = F(10;8) - F(9;8) = .099$.

    b. For a 30 minute period, $\alpha t = (4)(.5) = 2$, so $P(X = 0) = F(0;2) = .135$

    c. $E(X) = \alpha t = 2$

103. $\alpha = 1/(\text{mean time between occurrences}) = \dfrac{1}{.5} = 2$

    a. $\alpha t = (2)(2) = 4$

    b. $P(X > 5) 1 - P(X \leq 5) = 1 - .785 = .215$

    c. Solve for t, given $\alpha = 2$:
        $.1 = e^{-\alpha t}$
        $\ln(.1) = -\alpha t$
        $t = \dfrac{2.3026}{2} \approx 1.15$ years

105.

    a. For a one-quarter acre plot, the parameter is $(80)(.25) = 20$,
        so $P(X \leq 16) = F(16;20) = .221$

    b. The expected number of trees is $\alpha \cdot (\text{area}) = 80(85,000) = 6,800,000$.

    c. The area of the circle is $\pi r^2 = .031416$ sq. miles or 20.106 acres. Thus X has a
        Poisson distribution with parameter $\lambda = \alpha(20.106) = 1608.5$.

Chapter 3: Discrete Random Variables and Probability Distributions

107.

a. No events in $(0, t+\Delta t)$ if and only if no events in $(0, t)$ and no events in $(t, t+\Delta t)$.
Thus, $P_0(t+\Delta t) = P_0(t) \cdot P(\text{no events in } (t, t+\Delta t))$
$= P_0(t)[1 - \alpha \cdot \Delta t - o(\Delta t)]$

b. $\dfrac{P_0(t + \Delta t) - P_0(t)}{\Delta t} = -\alpha P_0(t) - P_0(t) \cdot \dfrac{o(\Delta t)}{\Delta t}$

c. $\dfrac{d}{dt}\left[e^{-\alpha t}\right] = -\alpha e^{-\alpha t} = -\alpha P_0(t)$, as desired.

d. $\dfrac{d}{dt}\left[\dfrac{e^{-\alpha t}(\alpha t)^k}{k!}\right] = \dfrac{-\alpha e^{-\alpha t}(\alpha t)^k}{k!} + \dfrac{k\alpha e^{-\alpha t}(\alpha t)^{k-1}}{k!}$

$= -\alpha\dfrac{e^{-\alpha t}(\alpha t)^k}{k!} + \alpha\dfrac{e^{-\alpha t}(\alpha t)^{k-1}}{(k-1)!} = -\alpha P_k(t) + \alpha P_{k-1}(t)$ as desired.

109. Re-write the binomial MGF as $(pe^t + 1 - p)^n = (1 + p(e^t - 1))^n = \left(1 + \dfrac{np(e^t - 1)}{n}\right)^n$.

Define $a_n = np(e^t - 1)$. From calculus, since $a_n \to \lambda(e^t - 1)$ by assumption, the expression above converges to $\exp(\lambda(e^t - 1))$, which is the Poisson MGF.

111.

a. $p(1) = P(\text{exactly one suit}) = P(\text{all spades}) + P(\text{all hearts}) + P(\text{all diamonds})$

$+ P(\text{all clubs}) = 4P(\text{all spades}) = 4 \cdot \dfrac{\dbinom{13}{5}}{\dbinom{52}{5}} = .00198$

$p(2) = P(\text{all hearts and spades with at least one of each}) + \ldots + P(\text{all diamonds and clubs with at least one of each})$
$= 6\, P(\text{all hearts and spades with at least one of each})$
$= 6\, [\, P(1\text{ h and }4\text{ s}) + P(2\text{ h and }3\text{ s}) + P(3\text{ h and }2\text{ s}) + P(4\text{ h and }1\text{ s})\,]$

$= 6 \cdot \left[\, 2 \cdot \dfrac{\dbinom{13}{4}\dbinom{13}{1}}{\dbinom{52}{5}} + 2 \cdot \dfrac{\dbinom{13}{3}\dbinom{13}{2}}{\dbinom{52}{5}}\, \right] = 6\left[\dfrac{18,590 + 44,616}{2,598,960}\right] = .14592$

Chapter 3: Discrete Random Variables and Probability Distributions

$$p(4) = 4P(2 \text{ spades, 1 h, 1 d, 1 c}) = \frac{4 \cdot \binom{13}{2}(13)(13)(13)}{\binom{52}{5}} = .26375$$

$$p(3) = 1 - [p(1) + p(2) + p(4)] = .58835$$

b. $\mu = \sum_{x=1}^{4} x \cdot p(x) = 3.114, \; \sigma^2 = \left[\sum_{x=1}^{4} x^2 \cdot p(x) \right] - (3.114)^2 = .405, \sigma = .636$

113.

    a. b(x;15,.75)

    b. P(X > 10) = 1 - B(10;15, .75) = 1 - .314 = .686

    c. B(10;15, .75) - B(5;15, .75) = .314 - .001 = .313

    d. $\mu = (15)(.75) = 11.75, \sigma^2 = (15)(.75)(.25) = 2.81$

    e. Requests can all be met if and only if $X \le 10$, and $15 - X \le 8$, i.e. if $7 \le X \le 10$, so P(all requests met) = B(10; 15,.75) - B(6; 15,.75) = .310

115. Let $X \sim \text{Bin}(5, .9)$. Then $P(X \ge 3) = 1 - P(X \le 2) = 1 - B(2;5,.9) = .991$

117.

    a. N = 500, p = .005, so np = 2.5 and b(x; 500, .005) $\approx$ p(x; 2.5), a Poisson p.m.f.

    b. P(X = 5) = p(5; 2.5) - p(4; 2.5) = .9580 - .8912 = .0668

    c. P(X $\ge$ 5) = 1 - p(4;2.5) = 1 - .8912 = .1088

119. Let Y denote the number of tests carried out. For n = 3, possible Y values are 1 and 4. $P(Y = 1) = P(\text{no one has the disease}) = (.9)^3 = .729$ and P(Y = 4) = .271, so E(Y) = (1)(.729) + (4)(.271) = 1.813, as contrasted with the 3 tests necessary without group testing. For n=5, possible values of Y are 1 and 6. $P(Y = 1) = (.9)^5 = .5905$, so P(Y = 6) = .4095 and E(Y) = (1)(.5905) + (6)(.4095) = 3.0475, less than the 5 tests necessary without group testing.

121. $p(2) = P(X = 2) = P(S \text{ on } \#1 \text{ and } S \text{ on } \#2) = p^2$
$p(3) = P(S \text{ on } \#3 \text{ and } S \text{ on } \#2 \text{ and } F \text{ on } \#1) = (1 - p)p^2$
$p(4) = P(S \text{ on } \#4 \text{ and } S \text{ on } \#3 \text{ and } F \text{ on } \#2) = (1 - p)p^2$
$p(5) = P(S \text{ on } \#5 \text{ and } S \text{ on } \#4 \text{ and } F \text{ on } \#3 \text{ and no 2 consecutive } S\text{'s on trials prior to } \#3) = [1 - p(2)](1 - p)p^2$

p(6) = P(S on #6 and S on #5 and F on #4 and no 2 consecutive S's on trials prior to #4) = [1 – p(2) – p(3)](1 – p)p²

In general, for x = 5, 6, 7, …: p(x) = [1 – p(2) - … – p(x - 3)](1 – p)p²

For p = .9,

x	2	3	4	5	6	7	8
p(x)	.81	.081	.081	.0154	.0088	.0023	.0010

123.

a. Let event C = seed carries single spikelets, and event P = seed produces ears with single spikelets. Then P(P ∩ C) = P(P | C) · P(C) = .29 (.40) = .116. Let X = the number of seeds out of the 10 selected that meet the condition P ∩ C. Then X ~ Bin(10, .116). $P(X = 5) = \binom{10}{5}(.116)^5(.884)^5 = .002857$

b. For 1 seed, the event of interest is P = seed produces ears with single spikelets.
P(P) = P(P ∩ C) + P(P ∩ C') = .116 (from **a**) + P(P | C') · P(C')
= .116 + (.26)(.60) = .272.
Let Y = the number out of the 10 seeds that meet condition P.
Then Y ~ Bin(10, .272), and P(Y = 5) = .0767.
P(Y ≤ 5) = b(0;10,.272) + … + b(5;10,.272) = .041813 + … + .076719 = .97024

125.

a. P(X = 0) = F(0;2) = 0.135

b. Let S = an operator who receives no requests. Then p = .135 and we wish P(4 S's in 5 trials) = b(4;5,.135) = $\binom{5}{4}(.135)^4(.865)^1$ = .00144

c. P(all receive x) = P(first receives x) · … · P(fifth receives x) = $\left[\dfrac{e^{-2}2^x}{x!}\right]^5$, and

P(all receive the same number) is the sum from x = 0 to ∞.

127. The number sold is min (X, 5), so E[min(x, 5)] = $\sum\limits^{\infty} \min(x,5)p(x;4)$ = (0)p(0;4) +

(1) p(1;4) + (2) p(2;4) + (3) p(3;4) + (4) p(4;4) + $5\sum\limits_{x=5}^{\infty} p(x;4)$ = 1.735 + 5[1 – F(4;4)]

= 3.59

Chapter 3: Discrete Random Variables and Probability Distributions

129.

a. No; probability of success is not the same for all tests.

b. There are four ways exactly three could have positive results. Let D represent those with the disease and D′ represent those without the disease.

	Combination		Probability
	D	**D′**	

D = 0, D′ = 3:

$$\left[\binom{5}{0}(.2)^0(.8)^5\right]\cdot\left[\binom{5}{3}(.9)^3(.1)^2\right]$$
$$=(.32768)(.0729)=.02389$$

D = 1, D′ = 2:

$$\left[\binom{5}{1}(.2)^1(.8)^4\right]\cdot\left[\binom{5}{2}(.9)2(.1)^3\right]$$
$$=(.4096)(.0081)=.00332$$

D = 2, D′ = 1:

$$\left[\binom{5}{2}(.2)^2(.8)^3\right]\cdot\left[\binom{5}{1}(.9)^1(.1)^4\right]$$
$$=(.2048)(.00045)=.00009216$$

D = 3, D′ = 0:

$$\left[\binom{5}{3}(.2)^3(.8)^2\right]\cdot\left[\binom{5}{0}(.9)^0(.1)^5\right]$$
$$=(.0512)(.00001)=.000000512$$

Adding up the probabilities associated with the four combinations yields 0.0273.

131.

a. $p(x;\lambda,\mu) = \frac{1}{2}p(x;\lambda) + \frac{1}{2}p(x;\mu)$ where both $p(x;\lambda)$ and $p(x;\mu)$ are Poisson p.m.f.'s and thus ≥ 0, so $p(x;\lambda,\mu) \geq 0$. Further,

$$\sum_{x=0}^{\infty}p(x;\lambda,\mu) = \frac{1}{2}\sum_{x=0}^{\infty}p(x;\lambda) + \frac{1}{2}\sum_{x=0}^{\infty}p(x;\mu) = \frac{1}{2}+\frac{1}{2}=1$$

b. $.6\,p(x;\lambda)+.4\,p(x;\mu)$

c. $E(X) = \sum_{x=0}^{\infty}x[\frac{1}{2}p(x;\lambda)+\frac{1}{2}p(x;\mu)] = \frac{1}{2}\sum_{x=0}^{\infty}xp(x;\lambda)+\frac{1}{2}\sum_{x=0}^{\infty}xp(x;\mu)$

$$= \frac{1}{2}\lambda + \frac{1}{2}\mu = \frac{\lambda + \mu}{2}$$

d. $E(X^2) = \frac{1}{2}\sum_{x=0}^{\infty} x^2 p(x;\lambda) + \frac{1}{2}\sum_{x=0}^{\infty} x^2 p(x;\mu) = \frac{1}{2}(\lambda^2 + \lambda) + \frac{1}{2}(\mu^2 + \mu)$ (since for a Poisson r.v., $E(X^2) = V(X) + [E(X)]^2 = \lambda + \lambda^2$),

so $V(X) = \frac{1}{2}[\lambda^2 + \lambda + \mu^2 + \mu] - \left[\frac{\lambda + \mu}{2}\right]^2 = \left(\frac{\lambda - \mu}{2}\right)^2 + \frac{\lambda + \mu}{2}$

133. $P(X = j) = \sum_{i=1}^{10} P\,(\text{arm on track } i \cap X = j) = \sum_{i=1}^{10} P\,(X = j \mid \text{arm on } i\,) \cdot p_i$

$$= \sum_{i=1}^{10} P\,(\text{next seek at I+j+1 or I-j-1}) \cdot p_i \;=\; \sum_{i=1}^{10}(p_{i+j+1} + P_{i-j-1})p_i$$

where $p_k = 0$ if $k < 0$ or $k > 10$

135. Let $A = \{x: |x - \mu| \geq k\sigma\}$. Then $\sigma^2 =$

$$\sum_{all\ x}(x - \mu)^2 p(x) \geq \sum_{A}(x - \mu)^2 p(x) \geq (k\sigma)^2 \sum_{A} p(x). \text{ But } \sum_{A} p(x) = P(X \text{ is in } A) =$$

$P(|X - \mu| \geq k\sigma)$, so $\sigma^2 \geq k^2\sigma^2 \cdot P(|X - \mu| \geq k\sigma)$, so $P(|X - \mu| \geq k\sigma) \leq 1/k^2$.

137. Assuming independence and constant probability, $X \sim \text{Bin}(25,p)$. Thus, $E[h(X)] = E[20X + 750] = 20E(X) + 750 = 20(25p) + 750 = 500p + 750$ and $SD[h(X)] = SD[20X + 750] = 20SD(X) = 20\sqrt{25pq} = 100\sqrt{pq}$.

Chapter 3: Discrete Random Variables and Probability Distributions

139. Let Y be the number of couples that arrive late, so $Y \sim \text{Bin}(3,.4)$, and let Z be the number of single individuals who arrive late, so $Z \sim \text{Bin}(2,.4)$. We are further given that Y and Z are independent, and we want the PMF of $X = 2Y + Z$. Exhaust all options for Y and Z, then collect terms.

y	z	$P(Y = y)$	$P(Z = z)$	$x = 2y+z$	$P(X = x)$
0	0	.216	.36	0	.07776
0	1	.216	.48	1	.10368
0	2	.216	.16	2	.03456
1	0	.432	.36	2	.15552
1	1	.432	.48	3	.20736
1	2	.432	.16	4	.06912
2	0	.288	.36	4	.10368
2	1	.288	.48	5	.13824
2	2	.288	.16	6	.04608
3	0	.064	.36	6	.02304
3	1	.064	.48	7	.03072
3	2	.064	.16	8	.01024

The probabilities for Y and Z are based on their binomial distributions, while the probabilities for X are the corresponding products (exploiting independence). The final PMF of X is

x	0	1	2	3	4	5	6	7	8
$p_X(x)$	.07776	.10368	.19008	.20736	.17280	.13824	.06912	.03072	.01024

Chapter 4: Continuous Random Variables and Probability Distributions

1.

    a. $P(X \leq 1) = \int_{-\infty}^{1} f(x)dx = \int_{0}^{1} \frac{1}{2}xdx = \frac{1}{4}x^2\Big]_{0}^{1} = .25$

    b. $P(.5 \leq X \leq 1.5) = \int_{.5}^{1.5} \frac{1}{2}xdx = \frac{1}{4}x^2\Big]_{.5}^{1.5} = .5$

    c. $P(X > 1.5) = \int_{.5}^{\infty} f(x)dx = \int_{.5}^{2} \frac{1}{2}xdx = \frac{1}{4}x^2\Big]_{1.5}^{2} = \frac{7}{16} \approx .438$

3.

    a. Graph of $f(x) = .09375(4 - x^2)$

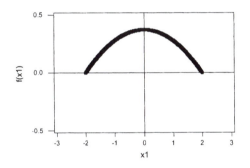

    b. $P(X > 0) = \int_{0}^{2} .09375(4 - x^2)dx = .09375(4x - \frac{x^3}{3})\Big]_{0}^{2} = .5$

    c. $P(-1 < X < 1) = \int_{-1}^{1} .09375(4 - x^2)dx = .6875$

    d. $P(x < -.5 \text{ OR } x > .5) = 1 - P(-.5 \leq X \leq .5) = 1 - \int_{-.5}^{.5} .09375(4 - x^2)dx$

                $= 1 - .3672 = .6328$

Chapter 4: Continuous Random Variables and Probability Distributions

5.

    a. $1 = \int_{-\infty}^{\infty} f(x)dx = \int_{0}^{2} kx^2 dx = k\left(\frac{x^3}{3}\right)\Big|_{0}^{2} = k\left(\frac{8}{3}\right) \Rightarrow k = \frac{3}{8}$

    b. $P(0 \le X \le 1) = \int_{0}^{1} \frac{3}{8} x^2 dx = \frac{1}{8} x^3 \Big|_{0}^{1} = \frac{1}{8} = .125$

    c. $P(1 \le X \le 1.5) = \int_{1}^{1.5} \frac{3}{8} x^2 dx = \frac{1}{8} x^3 \Big|_{1}^{1.5} = \frac{1}{8}\left(\frac{3}{2}\right)^3 - \frac{1}{8}(1)^3 = \frac{19}{64} \approx .2969$

    d. $P(X \ge 1.5) = 1 -$
$$\int_{0}^{1.5} \frac{3}{8} x^2 dx = \frac{1}{8} x^3 \Big|_{0}^{1.5} = 1 - \left[\frac{1}{8}\left(\frac{3}{2}\right)^3 - 0\right] = 1 - \frac{27}{64} = \frac{37}{64} \approx .5781$$

7.

    a. $f(x) = \frac{1}{10}$ for $25 \le x \le 35$ and $= 0$ otherwise

    b. $P(X > 33) = \int_{33}^{35} \frac{1}{10} dx = .2$

    c. $E(X) = \int_{25}^{35} x \cdot \frac{1}{10} dx = \frac{x^2}{20}\Big|_{25}^{35} = 30$

        30 ± 2 is from 28 to 32 minutes:

        $P(28 < X < 32) = \int_{28}^{32} \frac{1}{10} dx = \frac{1}{10} x \Big|_{28}^{32} = .4$

    d. $P(a \le x \le a+2) = \int_{a}^{a+2} \frac{1}{10} dx = .2$, since the interval has length 2.

9.

    a. $P(X \le 6) = = \int_{5}^{6} .15 e^{-.15(x-5)} dx = .15 \int_{0}^{6.5} e^{-.15u} du$ (after $u = x - .5$)

        $= e^{-.15u}\Big|_{0}^{5.5} = 1 - e^{-.825} \approx .562$

    b. $1 - .562 = .438; .438$

    c. $P(5 \le Y \le 6) = P(Y \le 6) - P(Y \le 5) \approx .562 - .491 = .071$

Chapter 4: Continuous Random Variables and Probability Distributions

11.

    a. $P(X \leq 1) = F(1) = \frac{1}{4} = .25$

    b. $P(.5 \leq X \leq 1) = F(1) - F(.5) = \frac{3}{16} = .1875$

    c. $P(X > .5) = 1 - P(X \leq .5) = 1 - F(.5) = \frac{15}{16} = .9375$

    d. $.5 = F(\tilde{\mu}) = \dfrac{\tilde{\mu}^2}{4} \Rightarrow \tilde{\mu}^2 = 2 \Rightarrow \tilde{\mu} = \sqrt{2} \approx 1.414$

    e. $f(x) = F'(x) = \frac{x}{2}$ for $0 \leq x < 2$, and $= 0$ otherwise

    f. $E(X) = \displaystyle\int_{-\infty}^{\infty} x \cdot f(x)dx = \int_{0}^{2} x \cdot \frac{1}{2}xdx = \frac{1}{2}\int_{0}^{2} x^2 dx = \left.\frac{x^3}{6}\right]_{0}^{2} = \frac{8}{6} \approx 1.333$

    g. $E(X^2) = \displaystyle\int_{-\infty}^{\infty} x^2 f(x)dx = \int_{0}^{2} x^2 \frac{1}{2}xdx = \frac{1}{2}\int_{0}^{2} x^3 dx = \left.\frac{x^4}{8}\right]_{0}^{2} = 2,$

        So $\text{Var}(X) = E(X^2) - [E(X)]^2 = 2 - \left(\frac{8}{6}\right)^2 = \frac{8}{36} \approx .222$, $\sigma_x \approx .471$

    h. From **g**, $E(X^2) = 2$

13.

    a. $1 = \displaystyle\int_{1}^{\infty} \frac{k}{x^4} dx \Rightarrow 1 = \left.\frac{-k}{3}x^{-3}\right|_{1}^{\infty} \Rightarrow 1 = 0 - (-\frac{k}{3})(1) \Rightarrow 1 = \frac{k}{3} \Rightarrow k = 3$

    b. cdf: $F(x) = \displaystyle\int_{-\infty}^{x} f(y)dy = \int_{1}^{x} 3y^{-4} dy = \left.-\frac{3}{3}y^{-3}\right|_{1}^{x} = -x^{-3} + 1 = 1 - \frac{1}{x^3}.$

        So $F(x) = \begin{cases} 0, & x \leq 1 \\ 1 - x^{-3}, & x > 1 \end{cases}$

    c. $P(x > 2) = 1 - F(2) = 1 - \left(1 - \frac{1}{8}\right) = \frac{1}{8}$ or $.125$;

        $P(2 < x < 3) = F(3) - F(2) = \left(1 - \frac{1}{27}\right) - \left(1 - \frac{1}{8}\right) = .963 - .875 = .088$

Chapter 4: Continuous Random Variables and Probability Distributions

15.

    a. $F(x) = 0$ for $x < 0$ and $F(x) = 1$ for $x > 2$. For $0 \le x \le 2$,

$$F(x) = \int_0^x \tfrac{3}{8} y^2 \, dy = \tfrac{1}{8} y^3 \Big]_0^x = \tfrac{1}{8} x^3$$

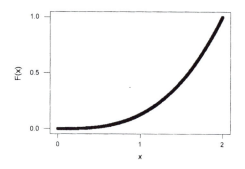

    b. $P(X \le .5) = F(.5) = \tfrac{1}{8} \left(\tfrac{1}{2}\right)^3 = \tfrac{1}{64}$

    c. $P(.25 \le X \le .5) = F(.5) - F(.25)$ $= \tfrac{1}{64} - \tfrac{1}{8}\left(\tfrac{1}{4}\right)^3 = \tfrac{7}{512} \approx .0137$

    d. $.75 = F(x) = \tfrac{1}{8} x^3 \Rightarrow x^3 = 6 \Rightarrow x \approx 1.8171$

17.

    a. $P(Y \le 1.8 \widetilde{\mu} + 32) = P(1.8X + 32 \le 1.8 \widetilde{\mu} + 32) = P(X \le \widetilde{\mu}) = .5$

    b. 90th for $Y = 1.8\eta(.9) + 32$ where $\eta(.9)$ is the 90th percentile for X, since
        $P(Y \le 1.8\eta(.9) + 32) = P(1.8X + 32 \le 1.8\eta(.9) + 32)$
                         $= (X \le \eta(.9)) = .9$ as desired.

    c. The $(100p)$th percentile for Y is $1.8\eta(p) + 32$, verified by substituting p for .9 in
        the argument of **b**. When $Y = aX + b$, (i.e. a linear transformation of X), and the
        $(100p)$th percentile of the X distribution is $\eta(p)$, then the corresponding $(100p)$th
        percentile of the Y distribution is $a\cdot\eta(p) + b$. (same linear transformation applied
        to X's percentile)

19.

a. $E(X) = \int_{-\infty}^{\infty} x\left(\frac{3}{x^4}\right) dx = \int_{1}^{\infty}\left(\frac{3}{x^3}\right) dx = -\frac{3}{2} x^{-2}\Big|_{1}^{\infty} = 0 + \frac{3}{2} = \frac{3}{2}$

  $E(X^2) = \int_{-\infty}^{\infty} x^2\left(\frac{3}{x^4}\right) dx = \int_{1}^{\infty}\left(\frac{3}{x^2}\right) dx = -3x^{-1}\Big|_{1}^{\infty} = 0 + 3 = 3$

  $V(X) = E(X^2) - [E(X)]^2 = 3 - \left(\frac{3}{2}\right)^2 = 3 - \frac{9}{4} = \frac{3}{4}$ or .75

  $\sigma = \sqrt{V(x)} = \sqrt{\frac{3}{4}} = .866$

b. $P(1.5 - .866 < x < 1.5 + .866) = P(x < 2.366) = F(2.366) = 1 - (2.366^{-3}) = .9245$

21.

a. $E(X) = \int_{-\infty}^{\infty} x \cdot f(x) dx = \int_{0}^{1} x \cdot 90x^8 (1-x) dx = 90 \int_{0}^{1} x^9 (1-x) dx$

  $= 9x^{10} - \frac{90}{11} x^{11}\Big|_{0}^{1} = \frac{9}{.11} \approx .8182$

  $E(X^2) = \int_{-\infty}^{\infty} x^2 \cdot f(x) dx = \int_{0}^{1} x^2 \cdot 90x^8 (1-x) dx = 90 \int_{0}^{1} x^{10} (1-x) dx$

  $= \frac{90}{11} x^{11} - \frac{90}{12} x^{12}\Big|_{0}^{1} \approx .6818$

  $V(X) \approx .6818 - (.8182)^2 = .0124, \qquad \sigma_x = .11134.$

b. $\mu \pm 2\sigma = (.5955, 1.0409)$. Thus, $P(\mu - 2\sigma \le X \le \mu + 2\sigma) = F(1.0409) - F(.5955) = 1 - .0437 = .9563$, so the probability X is *more* than 2 sd from its mean equals .0437.

23.

a. $F(X) = \frac{x - A}{B - A} = p \quad \Rightarrow \quad x = (100p)\text{th percentile} = A + (B - A)p$

b. $E(X) = \int_{A}^{B} x \cdot \frac{1}{B - A} dx = \frac{1}{B - A} \cdot \frac{x^2}{2}\Big|_{A}^{B} = \frac{1}{2} \cdot \frac{1}{B - A} \cdot (B^2 - A^2) = \frac{A + B}{2}$

  $E(X^2) = \frac{1}{3} \cdot \frac{1}{B - A} \cdot (B^3 - A^3) = \frac{A^2 + AB + B^2}{3}$

Chapter 4: Continuous Random Variables and Probability Distributions

$$V(X) = \left(\frac{A^2 + AB + B^2}{3}\right) - \left(\frac{(A+B)}{2}\right)^2 = \frac{(B-A)^2}{12}, \quad \sigma_x = \frac{(B-A)}{\sqrt{12}}$$

c. $E(X^n) = \int_A^B x^n \cdot \frac{1}{B-A} dx = \frac{B^{n+1} - A^{n+1}}{(n+1)(B-A)}$

25. $E(\text{area}) = E(\pi R^2) = \int_{-\infty}^{\infty} \pi r^2 f(r) dr = \int_9^{11} \pi r^2 \left(\frac{3}{4}\right)\left(1 - (10-r)^2\right) dr = \frac{501}{5} \pi = 314.79$

27. With X = temperature in °C, temperature in °F $= \frac{9}{5}X + 32,$ so

$$E\left[\frac{9}{5}X + 32\right] = \frac{9}{5}(120) + 32 = 248, \quad Var\left[\frac{9}{5}X + 32\right] = \left(\frac{9}{5}\right)^2 \cdot (2)^2 = 12.96, \text{ so } \sigma = 3.6$$

29. $M_X(t) = E[e^{tX}] = \int_0^{\infty} e^{tx} 4e^{-4x} dx = 4\int_0^{\infty} e^{-(4-t)x} dx = \frac{4}{4-t}$ for $t < 4$. From this, $E[X] =$

$M_X'(0) = \left.\frac{4}{(4-t)^2}\right|_{t=0} = .25$, and $E[X^2] = M_X''(0) = \left.\frac{2 \cdot 4}{(4-t)^3}\right|_{t=0} = .125 \rightarrow V(X) =$

$.125 - (.25)^2 = .0625.$

31. $M_X(0) = 1$; this is true for any MGF. For $t \neq 0$, $M_X(t) = \int_A^B e^{tx} \frac{1}{B-A} dx =$

$\left.\frac{1}{B-A} \frac{1}{t} e^{tx}\right|_A^B = \frac{e^{Bt} - e^{At}}{(B-A)t}.$

33. $M_X(t) = \int_{-\infty}^{\infty} e^{tx} .5 e^{-|x|} dx = .5\int_{-\infty}^0 e^{tx+x} dx + .5\int_0^{\infty} e^{tx-x} dx = \left.\frac{.5}{t+1} e^{(t+1)x}\right|_{-\infty}^0 + \left.\frac{.5}{t-1} e^{(t-1)x}\right|_0^{\infty} =$

$\frac{.5}{t+1} - \frac{.5}{t-1} = \frac{1}{1-t^2}$. The two integrals above converge only if $(t+1) > 0$ and if $(t-1) <$

0. These restrictions are identical to saying $|t| < 1$.

35. $R_X(t) = \ln[M_X(t)] = \ln(.15) + .5t - \ln(.15-t)$. Thus, $\mu = R_X'(0) = .5 + \frac{1}{.15-0} = 7.1667$

and $\sigma^2 = R_X''(0) = \frac{1}{(.15-0)^2} = 44.444.$

Chapter 4: Continuous Random Variables and Probability Distributions

37. $M_X(t) = \int_0^\infty e^{tx} .15 e^{-.15x} dx = .15 \int_0^\infty e^{-(.15-t)x} dx = \dfrac{.15}{.15-t}$. From this, $E[X] = M'_X(0) =$

$\dfrac{.15}{(.15-t)^2}\Big|_{t=0} = 6.6667$, and $E[X^2] = M''_X(0) = \dfrac{2(.15)}{(.15-t)^3}\Big|_{t=0} = 88.8889 \rightarrow V(X) =$

44.444. The distributions in Exercises 34 and 37 are just translations of each other: the PDF in Exercise 34 is the Exponential PDF of Exercise 37 shifted .5 units to the right. Hence, they have nearly the same MGFs, their means differ by .5, and they have the same variance.

39.

    a. $P(0 \le Z \le 2.17) = \Phi(2.17) - \Phi(0) = .4850$

    b. $\Phi(1) - \Phi(0) = .3413$

    c. $\Phi(0) - \Phi(-2.50) = .4938$

    d. $\Phi(2.50) - \Phi(-2.50) = .9876$

    e. $\Phi(1.37) = .9147$

    f. $P(-1.75 < Z) + [1 - P(Z < -1.75)] = 1 - \Phi(-1.75) = .9599$

    g. $\Phi(2) - \Phi(-1.50) = .9104$

    h. $\Phi(2.50) - \Phi(1.37) = .0791$

    i. $1 - \Phi(1.50) = .0668$

    j. $P(|Z| \le 2.50) = P(-2.50 \le Z \le 2.50) = \Phi(2.50) - \Phi(-2.50) = .9876$

41.

    a. $\Phi(c) = .9100 \Rightarrow c \approx 1.34$ (.9099 is the entry in the 1.3 row, .04 column)

    b. 9th percentile = -91st percentile = -1.34

    c. $\Phi(c) = .7500 \Rightarrow c \approx .675$ since .7486 and .7517 are in the .67 and .68 entries, respectively.

    d. $25^{th} = -75^{th} = -.675$

    e. $\Phi(c) = .06 \Rightarrow c \approx -1.555$ (.0594 and .0606 appear as the -1.56 and -1.55 entries, respectively).

Chapter 4: Continuous Random Variables and Probability Distributions

43.

    a. $P(X \le 100) = P\left(z \le \dfrac{100 - 80}{10}\right) = P(Z \le 2) = \Phi(2.00) = .9772$

    b. $P(X \le 80) = P\left(z \le \dfrac{80 - 80}{10}\right) = P(Z \le 0) = \Phi(0.00) = .5$

    c. $P(65 \le X \le 100) = P\left(\dfrac{65 - 80}{10} \le z \le \dfrac{100 - 80}{10}\right) = P(-1.50 \le Z \le 2)$

        $= \Phi(2.00) - \Phi(-1.50) = .9772 - .0668 = .9104$

    d. $P(70 \le X) = P(-1.00 \le Z) = 1 - \Phi(-1.00) = .8413$

    e. $P(85 \le X \le 95) = P(.50 \le Z \le 1.50) = \Phi(1.50) - \Phi(.50) = .2417$

    f. $P(|X - 80| \le 10) = P(-10 \le X - 80 \le 10) = P(70 \le X \le 90)$

                          $P(-1.00 \le Z \le 1.00) = .6826$

45.

    a. $P(X > .25) = P(Z > -.83) = 1 - .2033 = .7967$

    b. $P(X \le .10) = \Phi(-3.33) = .0004$

    c. We want the value of the distribution, c, that is the 95th percentile (5% of the values are higher). The 95th percentile of the standard normal distribution = 1.645. So c = .30 + (1.645)(.06) = .3987. The largest 5% of all concentration values are above .3987 mg/cm^3.

47.   Let X denote the diameter of a randomly selected cork made by the first machine, and let Y be defined analogously for the second machine.

    $P(2.9 \le X \le 3.1) = P(-1.00 \le Z \le 1.00) = .6826$

    $P(2.9 \le Y \le 3.1) = P(-7.00 \le Z \le 3.00) = .9987$

    So the second machine wins handily.

49. $\mu = 43; \sigma = 4.5$

    a. $P(X < 40) = P\left(z \le \dfrac{40 - 43}{4.5}\right) = P(Z < -0.667) = .2514$

        $P(X > 60) = P\left(z > \dfrac{60 - 43}{4.5}\right) = P(Z > 3.778) \approx 0$

    b. $43 + (-0.67)(4.5) = 39.985$

Chapter 4: Continuous Random Variables and Probability Distributions

51. From Table A.3, $P(-1.96 \leq Z \leq 1.96) = .95$. Then $P(\mu - .1 \leq X \leq \mu + .1) =$

$$P\left(\frac{-.1}{\sigma} < z < \frac{.1}{\sigma}\right) \text{ implies that } \frac{.1}{\sigma} = 1.96, \text{ and thus that } \sigma = \frac{.1}{1.96} = .0510$$

53.

    a. $P(\mu - 1.5\sigma \leq X \leq \mu + 1.5\sigma) = P(-1.5 \leq Z \leq 1.5) = \Phi(1.50) - \Phi(-1.50) = .8664$

    b. $P(X < \mu - 2.5\sigma \text{ or } X > \mu + 2.5\sigma) = 1 - P(\mu - 2.5\sigma \leq X \leq \mu + 2.5\sigma)$
        $= 1 - P(-2.5 \leq Z \leq 2.5) = 1 - .9876 = .0124$

    c. $P(\mu - 2\sigma \leq X \leq \mu - \sigma \text{ or } \mu + \sigma \leq X \leq \mu + 2\sigma) = P(\text{within 2 sd's}) - P(\text{within 1 sd})$
        $= P(\mu - 2\sigma \leq X \leq \mu + 2\sigma) - P(\mu - \sigma \leq X \leq \mu + \sigma)$
        $= .9544 - .6826 = .2718$

55.

    a. $P(67 \leq X \leq 75) = P(-1.00 \leq Z \leq 1.67) = .7938$

    b. $P(70 - c \leq X \leq 70 + c) = P\left(\frac{-c}{3} \leq Z \leq \frac{c}{3}\right) = 2\Phi(\frac{c}{3}) - 1 = .95 \Rightarrow \Phi(\frac{c}{3}) = .9750$

        $\frac{c}{3} = 1.96 \Rightarrow c = 5.88$

    c. $10 \cdot P(\text{a single one is acceptable}) = 7.938$

    d. $p = P(X < 73.84) = P(Z < 1.28) = .9$, so $P(Y \leq 8) = B(8;10,.9) = .264$

57. No, just use the symmetry of the Z curve about 0. That is, $\Phi(-z) = 1 - \Phi(z)$.

59. We use a Normal approximation to the Binomial distribution: $X \sim b(x;1000,.03) \approx$ $N(30,5.394)$.

    a. $P(x \geq 40) = 1 - P(x \leq 39) = 1 - P\left(Z \leq \frac{39.5 - 30}{5.394}\right)$
        $= 1 - \Phi(1.76) = 1 - .9608 = .0392$

    b. 5% of 1000 = 50: $P(x \leq 50) = P\left(Z \leq \frac{50.5 - 30}{5.394}\right) = \Phi(3.80) \approx 1.00$

61.

    a. $P(20 - .5 \leq X \leq 30 + .5) = P(19.5 \leq X \leq 30.5) = P(-1.1 \leq Z \leq 1.1) = .7286$

    b. $P(\text{at most } 30) = P(X \leq 30 + .5) = P(Z \leq 1.1) = .8643$.
        $P(\text{less than } 30) = P(X < 30 - .5) = P(Z < .9) = .8159$

Chapter 4: Continuous Random Variables and Probability Distributions

63. $p = .10$; $n = 200$; $np = 20$, $npq = 18$

a. $P(X \le 30) = \Phi\left(\dfrac{30 + .5 - 20}{\sqrt{18}}\right) = \Phi(2.47) = .9932$

b. $P(X < 30) = P(X \le 29) = \Phi\left(\dfrac{29 + .5 - 20}{\sqrt{18}}\right) = \Phi(2.24) = .9875$

c. $P(15 \le X \le 25) = P(X \le 25) - P(X \le 14) = \Phi\left(\dfrac{25 + .5 - 20}{\sqrt{18}}\right) - \Phi\left(\dfrac{14 + .5 - 20}{\sqrt{18}}\right)$

$\Phi(1.30) - \Phi(-1.30) = .9032 - .0968 = .8064$

65. $P(X \le \mu + \sigma[(100p)\text{th} \text{ percentile for std normal}])$

$P\left(\dfrac{X - \mu}{\sigma} \le [\ldots]\right) = P(Z \le [\ldots]) = p$ as desired

67.

a. $P(Z \ge 1) \approx .5 \cdot \exp\left(\dfrac{83 + 351 + 562}{703 + 165}\right) = .1587$

b. $P(Z > 3) \approx .5 \cdot \exp\left(\dfrac{-2362}{399.3333}\right) = .0013$

c. $P(Z > 4) \approx .5 \cdot \exp\left(\dfrac{-3294}{340.75}\right) = .0000317$, so $P(-4 < Z < 4) \approx 1 - 2(.0000317) = .999937$

d. $P(Z > 5) \approx .5 \cdot \exp\left(\dfrac{-4392}{305.6}\right) = .00000029$

69.

a. $\Gamma(6) = 5! = 120$

b. $\Gamma\left(\dfrac{5}{2}\right) = \dfrac{3}{2}\Gamma\left(\dfrac{1}{2}\right) = \dfrac{3}{2} \cdot \dfrac{1}{2} \cdot \Gamma\left(\dfrac{1}{2}\right) = \left(\dfrac{3}{4}\right)\sqrt{\pi} \approx 1.329$

c. $F(4;5) = .371$ from row 4, column 5 of Table A.4

    d. $F(5;4) = .735$

    e. $F(0;4) = P(X \le 0; \alpha = 4) = 0$

71.

    a. $\mu = 20, \ \sigma^2 = 80 \Rightarrow \alpha\beta = 20, \ \alpha\beta^2 = 80 \Rightarrow \beta = \frac{80}{20}, \ \alpha = 5$

    b. $P(X \le 24) = F\left(\frac{24}{4};5\right) = F(6;5) = .715$

    c. $P(20 \le X \le 40) = F(10;5) - F(5;5) = .411$

73.

    a. $E(X) = \dfrac{1}{\lambda} = 1$

    b. $\sigma = \dfrac{1}{\lambda} = 1$

    c. $P(X \le 4) = 1 - e^{-(1)(4)} = 1 - e^{-4} = .982$

    d. $P(2 \le X \le 5) = 1 - e^{-(1)(5)} - \left[1 - e^{-(1)(2)}\right] = e^{-2} - e^{-5} = .129$

75.   Mean $= \dfrac{1}{\lambda} = 25{,}000$ implies $\lambda = .00004$

    a. $P(X > 20{,}000) = 1 - P(X \le 20{,}000) = 1 - F(20{,}000; .00004)$
       $= e^{-(.00004)(20{,}000)} = .449$
       $P(X \le 30{,}000) = F(30{,}000; .00004) = 1 - e^{-1.2} = .699$
       $P(20{,}000 \le X \le 30{,}000) = .699 - .551 = .148$

    b. $\sigma = \dfrac{1}{\lambda} = 25{,}000$, so $P(X > \mu + 2\sigma) = P(X > 75{,}000) =$
           $1 - F(75{,}000; .00004) = .05.$
       Similarly, $P(X > \mu + 3\sigma) = P(X > 100{,}000) = .018$

Chapter 4: Continuous Random Variables and Probability Distributions

77.

    a. $\{X \geq t\} = A_1 \cap A_2 \cap A_3 \cap A_4 \cap A_5$

    b. $P(X \geq t) = P(A_1) \cdot P(A_2) \cdot P(A_3) \cdot P(A_4) \cdot P(A_5) = \left(e^{-\lambda t}\right)^5 = e^{-.05t}$, so $F_x(t)$

        $= P(X \leq t) = 1 - e^{-.05t}$, $f_x(t) = .05e^{-.05t}$ for $t \geq 0$. Thus X also ha an exponential distribution, but with parameter $\lambda = .05$.

    c.   By the same reasoning, $P(X \leq t) = 1 - e^{-n\lambda t}$, so X has an exponential distribution with parameter $n\lambda$.

79.

    a. $\{X^2 \leq y\} = \{-\sqrt{y} \leq X \leq \sqrt{y}\}$

    b. $F_Y(y) = P(X^2 \leq y) = \int_{-\sqrt{y}}^{\sqrt{y}} \frac{1}{\sqrt{2\pi}} e^{-z^2/2} dz$, so $f_Y(y) =$

        $\frac{1}{\sqrt{2\pi}} e^{-y/2} \frac{1}{2} y^{-1/2} - \frac{1}{\sqrt{2\pi}} e^{-y/2} \frac{-1}{2} y^{-1/2} = \frac{1}{\sqrt{2\pi}} e^{-y/2} y^{-1/2}$. We recognize this as the chi-squared p.d.f. with $\nu = 1$.

81.   The gamma density function implies that $\int x^{a-1} e^{-x/\beta} dx = \beta^a \Gamma(a)$ for any $a, \beta > 0$.

We will use this twice, once with $a = \alpha + 1$ and once with $a = \alpha + 2$. We will also use the property $\Gamma(a+1) = a\Gamma(a)$ for any $a > 0$.

First, $E[X] = \int_0^\infty x \frac{1}{\beta^\alpha \Gamma(\alpha)} x^{\alpha-1} e^{-x/\beta} dx = \frac{1}{\beta^\alpha \Gamma(\alpha)} \int_0^\infty x^{(\alpha+1)-1} e^{-x/\beta} dx =$

$\frac{1}{\beta^\alpha \Gamma(\alpha)} \beta^{\alpha+1} \Gamma(\alpha+1) = \frac{1}{\Gamma(\alpha)} \beta \alpha \Gamma(\alpha) = \alpha\beta$. Second, $E[X^2] =$

$\int_0^\infty x^2 \frac{1}{\beta^\alpha \Gamma(\alpha)} x^{\alpha-1} e^{-x/\beta} dx = \frac{1}{\beta^\alpha \Gamma(\alpha)} \int_0^\infty x^{(\alpha+2)-1} e^{-x/\beta} dx = \frac{1}{\beta^\alpha \Gamma(\alpha)} \beta^{\alpha+2} \Gamma(\alpha+2)$

$= \frac{1}{\Gamma(\alpha)} \beta^2 (\alpha+1) \Gamma(\alpha+1) = \frac{1}{\Gamma(\alpha)} \beta^2 (\alpha+1) \alpha \Gamma(\alpha) = \beta^2 \alpha(\alpha+1) \rightarrow V(X) = \beta^2 \alpha(\alpha+1) - (\alpha\beta)^2 = \alpha\beta^2$.

Chapter 4: Continuous Random Variables and Probability Distributions

83.

   a. $P(X \le 250) = F(250;2.5, 200) = 1 - e^{-(250/200)^{2.5}} = 1 - e^{-1.75} \approx .8257$

      $P(X < 250) = P(X \le 250) \approx .8257$

      $P(X > 300) = 1 - F(300; 2.5, 200) = e^{-(1.5)^{2.5}} = .0636$

   b. $P(100 \le X \le 250) = F(250;2.5, 200) - F(100;2.5, 200) \approx .8257 - .162 = .6637$

   c. The median $\tilde{\mu}$ is requested. The equation $F(\tilde{\mu}) = .5$ reduces to

      $.5 = e^{-(\tilde{\mu}/200)^{2.5}}$, i.e., $\ln(.5) \approx -\left(\dfrac{\tilde{\mu}}{200}\right)^{2.5}$, so $\tilde{\mu} = (.6931)^{.4}(200) = 172.727$.

85. $\mu = \displaystyle\int_0^\infty x \cdot \frac{\alpha}{\beta^\alpha} x^{\alpha-1} e^{-(x/\beta)^\alpha} dx$ = [after $y = \left(\dfrac{x}{\beta}\right)^\alpha$, $dy = \dfrac{\alpha x^{\alpha-1}}{\beta^\alpha} dx$]

      $\beta \displaystyle\int_0^\infty y^{1/\alpha} e^{-y} dy = \beta \cdot \Gamma\left(1 + \dfrac{1}{\alpha}\right)$ by definition of the gamma function.

87. $X \sim$ Weibull: $\alpha=20, \beta=100$

   a. $F(x, 20, \beta) = 1 - e^{-\left(\frac{x}{\beta}\right)^r} = 1 - e^{-\left(\frac{105}{100}\right)^{20}} = 1 - .070 = .930$

   b. $F(105) - F(100) = .930 - \left(1 - e^{-1}\right) = .930 - .632 = .298$

   c. $.50 = 1 - e^{-\left(\frac{x}{100}\right)^{20}} \Rightarrow e^{-\left(\frac{x}{100}\right)^{20}} = .50 \Rightarrow -\left(\dfrac{x}{100}\right)^{20} = \ln(.50)$

      $\left(\dfrac{-x}{100}\right) = \sqrt[20]{\ln(.50)} \Rightarrow -x = 100\left(\sqrt[20]{\ln(.50)}\right) \Rightarrow x = 98.18$

89.

   a. $E(X) = e^{3.5 + (1.2)^2/2} = 68.0335$; $V(X) = e^{2(3.5) + (1.2)^2} \cdot \left(e^{(1.2)^2} - 1\right) = 14907.168$;

      $\sigma_x = 122.0949$

   b. $P(50 \le X \le 250) = P\left(z \le \dfrac{\ln(250) - 3.5}{1.2}\right) - P\left(z \le \dfrac{\ln(50) - 3.5}{1.2}\right)$

      $P(Z \le 1.68) - P(Z \le .34) = .9535 - .6331 = .3204.$

   c. $P(X \le 68.0335) = P\left(z \le \dfrac{\ln(68.0335) - 3.5}{1.2}\right) = P(Z \le .60) = .7257.$ The

      lognormal distribution is not a symmetric distribution.

Chapter 4: Continuous Random Variables and Probability Distributions

91.

   a. $E(X) = e^{5+(.01)/2} = e^{5.005} = 149.157$; $Var(X) = e^{10+(.01)} \cdot (e^{.01} - 1) = 223.594$

   b. $P(X > 125) = 1 - P(X \le 125) = = 1 - P\left(z \le \dfrac{\ln(125) - 5}{.1}\right) = 1 - \Phi(-1.72) = .9573$

   c. $P(110 \le X \le 125) = \Phi(-1.72) - \Phi\left(\dfrac{\ln(110) - 5}{.1}\right) = .0427 - .0013 = .0414$

   d. $\tilde{\mu} = e^{5} = 148.41$

   e.   P(any particular one has X > 125) = .9573 $\Rightarrow$ expected # = 10(.9573) = 9.573

   f.   We wish the 5th percentile, which is $e^{5+(-1.645)(.1)} = 125.90$

93.   The point of symmetry must be $\frac{1}{2}$, so we require that $f\left(\frac{1}{2} - \mu\right) = f\left(\frac{1}{2} + \mu\right)$, i.e.,

$\left(\frac{1}{2} - \mu\right)^{\alpha-1}\left(\frac{1}{2} + \mu\right)^{\beta-1} = \left(\frac{1}{2} + \mu\right)^{\alpha-1}\left(\frac{1}{2} - \mu\right)^{\beta-1}$, which in turn implies that $\alpha = \beta$.

95.

   a. $E(X) = \displaystyle\int_{0}^{1} x \cdot \frac{\Gamma(\alpha + \beta)}{\Gamma(\alpha)\Gamma(\beta)} x^{\alpha-1}(1-x)^{\beta-1} dx = \frac{\Gamma(\alpha + \beta)}{\Gamma(\alpha)\Gamma(\beta)} \int_{0}^{1} x^{\alpha}(1-x)^{\beta-1} dx$

       $\dfrac{\Gamma(\alpha + \beta)}{\Gamma(\alpha)\Gamma(\beta)} \cdot \dfrac{\Gamma(\alpha + 1)\Gamma(\beta)}{\Gamma(\alpha + \beta + 1)} = \dfrac{\alpha\Gamma(\alpha)}{\Gamma(\alpha)\Gamma(\beta)} \cdot \dfrac{\Gamma(\alpha + \beta)}{(\alpha + \beta)\Gamma(\alpha + \beta)} = \dfrac{\alpha}{\alpha + \beta}$

   b. $E[(1 - X)^{m}] = \displaystyle\int_{0}^{1}(1 - x)^{m} \cdot \frac{\Gamma(\alpha + \beta)}{\Gamma(\alpha)\Gamma(\beta)} x^{\alpha-1}(1-x)^{\beta-1} dx$

       $= \dfrac{\Gamma(\alpha + \beta)}{\Gamma(\alpha)\Gamma(\beta)} \displaystyle\int_{0}^{1} x^{\alpha-1}(1-x)^{m+\beta-1} dx = \dfrac{\Gamma(\alpha + \beta) \cdot \Gamma(m + \beta)}{\Gamma(\alpha + \beta + m)\Gamma(\beta)}$

       For m = 1, $E(1 - X) = \dfrac{\beta}{\alpha + \beta}$.

97.   The given probability plot is quite linear, and thus it is quite plausible that the tension distribution is normal.

Chapter 4: Continuous Random Variables and Probability Distributions

99. The z percentile values are as follows: -1.86, -1.32, -1.01, -0.78, -0.58, -0.40, -0.24,-0.08, 0.08, 0.24, 0.40, 0.58, 0.78, 1.01, 1.30, and 1.86. The accompanying probability plot is reasonably straight, and thus it would be reasonable to use estimating methods that assume a normal population distribution.

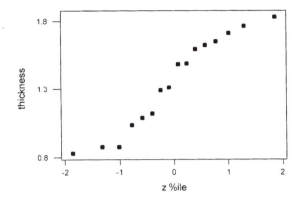

101. The (z percentile, observation) pairs are (-1.66, .736), (-1.32, .863), (-1.01, .865), (-.78, .913), (-.58, .915), (-.40, .937), (-.24, .983), (-.08, 1.007), (.08, 1.011), (.24, 1.064), (.40, 1.109), (.58, 1.132), (.78, 1.140), (1.01, 1.153), (1.32, 1.253), (1.86, 1.394). The accompanying probability plot is very straight, suggesting that an assumption of population normality is extremely plausible.

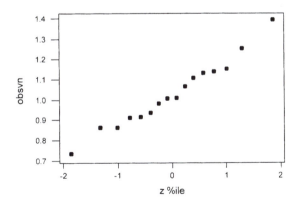

Chapter 4: Continuous Random Variables and Probability Distributions

103. To check for plausibility of a lognormal population distribution for the rainfall data of Exercise 81 in Chapter 1, take the natural logs and construct a normal probability plot. This plot and a normal probability plot for the original data appear below. Clearly the log transformation gives quite a straight plot, so lognormality is plausible. The curvature in the plot for the original data implies a positively skewed population distribution - like the lognormal distribution.

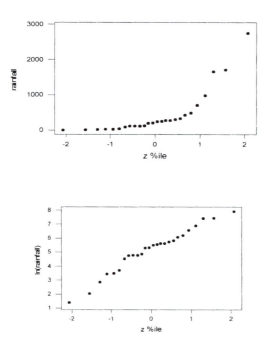

Chapter 4: Continuous Random Variables and Probability Distributions

105. The pattern in the plot (below, generated by Minitab) is reasonably linear. By visual inspection alone, it is plausible that strength is normally distributed.

Normal Probability Plot

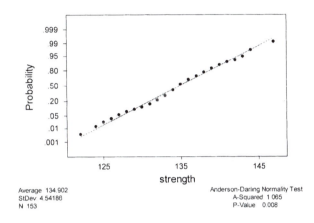

Average 134.902
StDev 4.54186
N 153

Anderson-Darling Normality Test
A-Squared 1.065
P-Value 0.008

Chapter 4: Continuous Random Variables and Probability Distributions

107. The $(100p)^{th}$ percentile $\eta(p)$ for the exponential distribution with $\lambda = 1$ satisfies $F(\eta(p)) = 1 - \exp[-\eta(p)] = p$, i.e., $\eta(p) = -\ln(1 - p)$. With $n = 16$, we need $\eta(p)$ for $p = \frac{5}{16}, \frac{1.5}{16}, ..., \frac{15.5}{16}$. These are .032, .398, .170, .247, .330, .421, .521, .633, .758, .901, 1.068, 1.269, 1.520, 1.856, 2.367, 3.466. this plot exhibits substantial curvature, casting doubt on the assumption of an exponential population distribution. Because λ is a scale parameter (as is σ for the normal family), $\lambda = 1$ can be used to assess the plausibility of the entire exponential familyThe $(100p)^{th}$ percentile $\eta(p)$ for the exponential distribution with $\lambda = 1$ satisfies $F(\eta(p)) = 1 - \exp[-\eta(p)] = p$, i.e., $\eta(p) = -\ln(1 - p)$. With $n = 16$, we need $\eta(p)$ for $p = \frac{5}{16}, \frac{1.5}{16}, ..., \frac{15.5}{16}$. These are .032, .398, .170, .247, .330, .421, .521, .633, .758, .901, 1.068, 1.269, 1.520, 1.856, 2.367, 3.466. this plot exhibits substantial curvature, casting doubt on the assumption of an exponential population distribution. Because λ is a scale parameter (as is σ for the normal family), $\lambda = 1$ can be used to assess the plausibility of the entire exponential family.

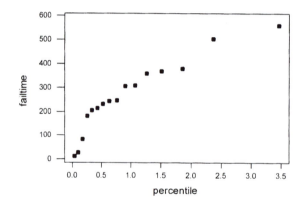

109. $y = 1/x \rightarrow x = 1/y$ and $0 < x < 1 \rightarrow 0 < 1/y < 1 \rightarrow y > 1$. Thus, $f_Y(y) = f_X(1/y)|dx/dy| = f_X(1/y)|-1/y^2| = 2(1/y)(1/y^2) = 2/y^3$ for $y > 1$.

111. $y = \sqrt{x} \rightarrow x = y^2$ and $x > 0 \rightarrow y > 0$. Thus, $f_Y(y) = f_X(y^2)|dx/dy| = (1/2)\exp(-y^2/2)|2y| = y\exp(-y^2/2)$ for $y > 0$.

113. $Y = X^2$, so $x = \sqrt{y}$ and $0 < x < 4 \rightarrow 0 < y < 16$. Thus, $f_Y(y) = f_X(\sqrt{y})|dx/dy| = \sqrt{y}/8|1/(2\sqrt{y})| = 1/16$ for $0 < y < 16$. That is, the area Y is uniform on $(0,16)$.

Chapter 4: Continuous Random Variables and Probability Distributions

115. $y = \tan(\pi(x-.5)) \rightarrow x = [\arctan(y)+.5]/\pi$ and $0 < x < 1 \rightarrow -\pi/2 < \pi(x-.5) < \pi/2 \rightarrow -\infty < y$
$< \infty$. From these, $f_Y(y) = f_X(x)|dx/dy| = 1\left|\dfrac{1}{\pi}\dfrac{1}{1+y^2}\right| = \dfrac{1}{\pi(1+y^2)}$ for $-\infty < y < \infty$.

117. We know that $F_X(X)$ is Uniform[0,1] for any continuous random variable. Here, $F_X(x) = x^2/16$ for $0 < x < 4$, and so $Y = g(X) = F_X(X) = X^2/16$

119. The transformation $y = x^2$ is not monotone on $[-1,1]$, so we must proceed via the CDF method. For $0 < y < 1$, $F_Y(y) = P(Y \le y) = P(X^2 \le y) = P(-\sqrt{y} \le X \le \sqrt{y}) = (\sqrt{y} - \sqrt{y})/(1-1) = 2\sqrt{y}/2 = \sqrt{y}$. [We've used the Uniform[-1,1] CDF here.] Thus, $f_Y(y) = d/dy(\sqrt{y}) = 1/(2\sqrt{y})$ for $0 < y < 1$.

121. The transformation $y = x^2$ is not monotone on $[-1,3]$, so we must proceed via the CDF method. For $0 < y < 1$, $F_Y(y) = P(Y \le y) = P(X^2 \le y) = P(-\sqrt{y} \le X \le \sqrt{y}) = (\sqrt{y} - \sqrt{y})/(3-1) = 2\sqrt{y}/4 = \sqrt{y}/2$. Thus, $f_Y(y) = d/dy(\sqrt{y}/2) = 1/(4\sqrt{y})$ for $0 < y < 1$.
For $1 < y < 9$, $F_Y(y) = P(Y \le y) = P(X^2 \le y) = P(-1 \le X \le 1$ or $1 < X \le \sqrt{y}) = \frac{1}{2} + (\sqrt{y}-1)/(3-1) = (1+\sqrt{y})/4$. Thus, $f_Y(y) = d/dy((1+\sqrt{y})/4) = 1/(8\sqrt{y})$ for $1 < y < 9$.
Together, $f_Y(y) = \begin{cases} 1/4\sqrt{y} & 0 < y < 1 \\ 1/8\sqrt{y} & 1 < y < 9 \end{cases}$

123. **a.** By assumption, the probability that you hit the disc centered at the bulls-eye with <u>area</u> x is proportional to x; in particular, this probability is $x/[$total area of target$] = x/[\pi(1)^2] = x/\pi$. Therefore, $F_X(x) = P(X \le x) = P($you hit disc centered at the bulls-eye with area $x) = x/\pi$. From this, $f_X(x) = d/dx[x/\pi] = 1/\pi$ for $0 < x < \pi$. That is, X is uniform on $(0, \pi)$.
b. $x = \pi y^2$ and $0 < x < \pi \rightarrow 0 < y < 1$. Thus, $f_Y(y) = f_X(\pi y^2)|dx/dy| = 1/\pi |2\pi y| = 2y$ for $0 < y < 1$.

125. These rvs are discrete: $x = 0, 1, 2, \ldots \rightarrow y = 1, 2, 3, \ldots$. For any positive integer y, $p_Y(y) = P(Y=y) = P(X+1 = y) = P(X = y-1) = (1-p)^{y-1}p$.

Chapter 4: Continuous Random Variables and Probability Distributions

127.

a. $P(10 \le X \le 20) = \dfrac{10}{25} = .4$

b. $P(X \ge 10) = P(10 \le X \le 25) = \dfrac{15}{25} = .6$

c. For $0 \le X \le 25$, $F(x) = \displaystyle\int_0^x \dfrac{1}{25}\,dy = \dfrac{x}{25}$. $F(x)=0$ for $x < 0$ and $= 1$ for $x > 25$.

d. $E(X) = \dfrac{(A + B)}{2} = \dfrac{(0 + 25)}{2} = 12.5$; $\text{Var}(X) = \dfrac{(B - A)^2}{12} = \dfrac{625}{12} = 52.083$, so $\sigma_x = 7.22$

129.

a. Clearly $f(x) \ge 0$. The c.d.f. is, for $x > 0$,

$$F(x) = \int_{-\infty}^x f(y)\,dy = \int_0^x \dfrac{32}{(y+4)^3}\,dy = -\dfrac{1}{2}\cdot\dfrac{32}{(y+4)^2}\Bigg]_0^x = 1 - \dfrac{16}{(x+4)^2}$$

($F(x) = 0$ for $x \le 0$.)

Since $F(\infty) = \displaystyle\int_{-\infty}^\infty f(y)\,dy = 1$, $f(x)$ is a legitimate pdf.

b. See above

c. $P(2 \le X \le 5) = F(5) - F(2) = 1 - \dfrac{16}{81} - \left(1 - \dfrac{16}{36}\right) = .247$

d. $E(x) = \displaystyle\int_{-\infty}^\infty x\cdot f(x)\,dx = \int_{-\infty}^\infty x\cdot\dfrac{32}{(x+4)^3}\,dx = \int_0^\infty (x + 4 - 4)\cdot\dfrac{32}{(x+4)^3}\,dx$

$$= \int_0^\infty \dfrac{32}{(x+4)^2}\,dx - 4\int_0^\infty \dfrac{32}{(x+4)^3}\,dx = 8 - 4 = 4$$

e. $E(\text{salvage value}) = = \displaystyle\int_0^\infty \dfrac{100}{x+4}\cdot\dfrac{32}{(x+4)^3}\,dx = 3200\int_0^\infty \dfrac{1}{(x+4)^4}\,dx = \dfrac{3200}{(3)(64)} = 16.67$

Chapter 4: Continuous Random Variables and Probability Distributions

131. $\mu = 40$ V; $\sigma = 1.5$ V

   a. $P(39 < X < 42) = \Phi\left(\dfrac{42 - 40}{1.5}\right) - \Phi\left(\dfrac{39 - 40}{1.5}\right)$

       $= \Phi(1.33) - \Phi(-.67) = .9082 - .2514 = .6568$

   b. We desire the 85th percentile: $40 + (1.04)(1.5) = 41.56$

   c. $P(X > 42) = 1 - P(X \le 42) = 1 \ -\Phi\left(\dfrac{42 - 40}{1.5}\right) = 1 - \Phi(1.33) = .0918$

       Let D represent the number of diodes out of 4 with voltage exceeding 42.

       $P(D \ge 1) = 1 - P(D = 0) = 1 - \binom{4}{0}(.0918)^{0}(.9082)^{4} = 1 - .6803 = .3197$

133.

   a. Let S = defective. Then $p = P(S) = .05$; $n = 250 \Rightarrow \mu = np = 12.5$, $\sigma = 3.446$.
       The random variable X = the number of defectives in the batch of 250. X ~
       Binomial. Since np = 12.5 ≥ 10, and nq = 237.5 ≥ 10, we can use the normal
       approximation.

       $P(X_{bin} \ge 25) \approx 1 \ -\Phi\left(\dfrac{24.5 - 12.5}{3.446}\right) = 1 - \Phi(3.48) = 1 - .9997 = .0003$

   b. $P(X_{bin} = 10) \approx P(X_{norm} \le 10.5) - P(X_{norm} \le 9.5)$
       $= \Phi(-.58) - \Phi(-.87) = .2810 - .1922 = .0888$

135.

   a. $F(x) = 0$ for $x < 1$ and $= 1$ for $x > 3$. For $1 \le x \le 3$, $F(x) = \displaystyle\int_{-\infty}^{x} f(y)dy$

       $= \displaystyle\int_{-\infty}^{1} 0\,dy + \int_{1}^{x} \dfrac{3}{2}\cdot\dfrac{1}{y^{2}}\,dy = 1.5\left(1 - \dfrac{1}{x}\right)$

   b. $P(X \le 2.5) = F(2.5) = 1.5(1 - .4) = .9$; $P(1.5 \le x \le 2.5) =$
       $F(2.5) - F(1.5) = .4$

   c. $E(X) = \ = \displaystyle\int_{1}^{3} x\cdot\dfrac{3}{2}\cdot\dfrac{1}{x^{2}}\,dx = \dfrac{3}{2}\int_{1}^{3}\dfrac{1}{x}\,dx = 1.5\ln(x)\big]_{1}^{3} = 1.648$

   d. $E(X^{2}) = \ = \displaystyle\int_{1}^{3} x^{2}\cdot\dfrac{3}{2}\cdot\dfrac{1}{x^{2}}\,dx = \dfrac{3}{2}\int_{1}^{3} dx = 3$, so $V(X) = E(X^{2}) - [E(X)]^{2} = .284$,

       $\sigma = .553$

Chapter 4: Continuous Random Variables and Probability Distributions

   e. $h(x) = \begin{cases} 0 & 1 \leq x \leq 1.5 \\ x - 1.5 & 1.5 \leq x \leq 2.5 \\ 1 & 2.5 \leq x \leq 3 \end{cases}$

   so $E[h(X)] = = \int_{.5}^{2.5}(x - 1.5) \cdot \frac{3}{2} \cdot \frac{1}{x^2}\,dx + \int_{2.5}^{3} 1 \cdot \frac{3}{2} \cdot \frac{1}{x^2}\,dx = .267$

137.

   a. $E(X) = \frac{1}{\lambda} = 1.075,\ \sigma = \frac{1}{\lambda} = 1.075$

   b. $P(3.0 < X) = 1 - P(X \leq 3.0) = 1 - F(3.0) = e^{-.93(3.0)} = .0614$
       $P(1.0 \leq X \leq 3.0) = F(3.0) - F(1.0) = .333$

   c.   The 90th percentile is requested; denoting it by c, we have
       $.9 = F(c) = 1 - e^{-(.93)c}$, whence $c = \dfrac{\ln(.1)}{(-.93)} = 2.476$

139.

   a. $E(cX) = cE(X) = \dfrac{c}{\lambda}$

   b. $E[c(1 - .5e^{ax})] = \int_0^\infty c(1 - .5e^{ax}) \cdot \lambda e^{-\lambda x}\,dx = \dfrac{c[.5\lambda - a]}{\lambda - a}$

141.

   a. $\int_{-\infty}^{\infty} f(x)\,dx = \int_{-\infty}^{0} .1e^{.2x}\,dx + \int_0^\infty .1e^{-.2x}\,dx = .5 + .5 = 1$

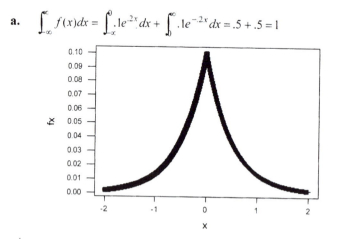

b. For x < 0, $F(x) = \int_{-\infty}^{x} .1e^{.2y} dy = \frac{1}{2} e^{.2x}$.

For x ≥ 0, $F(x) = \frac{1}{2} + \int_{0}^{x} .1e^{-.2y} dy = 1 - \frac{1}{2} e^{-.2x}$.

c. $P(X < 0) = F(0) = \frac{1}{2} = .5$, $P(X < 2) = F(2) = 1 - .5e^{-.4} = .665$,

$P(-1 \le X \le 2) - F(2) - F(-1) = .256$, $1 - (-2 \le X \le 2) = .670$

143.

a. $1 = \int_{5}^{\infty} \frac{k}{x^{\alpha}} dx = k \cdot \frac{5^{1-\alpha}}{\alpha - 1} \Rightarrow k = (\alpha - 1)5^{\alpha-1}$ where we must have α > 1.

b. For x ≥ 5, $F(x) = \int_{5}^{x} \frac{k}{y^{\alpha}} dy = 5^{1-\alpha}\left[\frac{1}{5^{1-\alpha}} - \frac{1}{x^{\alpha-1}} \right] = 1 - \left(\frac{5}{x} \right)^{\alpha-1}$.

c. $E(X) = \int_{5}^{\infty} x \cdot \frac{k}{x^{\alpha}} dx = \int_{5}^{\infty} \frac{(\alpha - 1)5^{\alpha-1}}{x^{\alpha-1}} dx = \frac{5(\alpha - 1)}{(\alpha - 2)}$, provided α > 2.

d. $P\left(\ln\left(\frac{X}{5} \right) \le y \right) = P\left(\frac{X}{5} \le e^{y} \right) = P(X \le 5e^{y}) = F(5e^{y}) = 1 - \left(\frac{5}{5e^{y}} \right)^{\alpha-1}$

$1 - e^{-(\alpha-1)y}$, the cdf of an exponential r.v. with parameter α - 1.

145.

a.

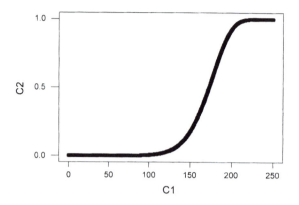

b. $P(X > 175) = 1 - F(175; 9, 180) = e^{-\left(\frac{175}{180}\right)^9} = .4602$
$P(150 \le X \le 175) = F(175; 9, 180) - F(150; 9, 180)$
$= .5398 - .1762 = .3636$

c. $P(\text{at least one}) = 1 - P(\text{none}) = 1 - (1 - .3636)^2 = .5950$

d. We want the 10th percentile: $.10 = F(x; 9, 180) = 1 - e^{-\left(\frac{x}{180}\right)^9}$. A small bit of algebra leads us to x = 140.178. Thus 10% of all tensile strengths will be less than 140.178 MPa.

147.

a. If we let $\alpha = 2$ and $\beta = \sqrt{2}\sigma$, then we can manipulate f(v) as follows:

$$f(v) = \frac{v}{\sigma^2} e^{-v^2/2\sigma^2} = \frac{2}{2\sigma^2} v e^{-v^2/2\sigma^2} = \frac{2}{\left(\sqrt{2}\sigma\right)^2} v^{2-1} e^{-\left(v/\sqrt{2}\sigma\right)^2} = \frac{\alpha}{\beta^\alpha} v^{\alpha-1} e^{-\left(v/\beta\right)^2},$$

which is in the Weibull family of distributions.

b. $F(v) = \int_0^{25} \frac{v}{400} e^{\frac{-v}{800}} dv$; cdf: $F\left(v; 2, \sqrt{2}\sigma\right) = 1 - e^{-\left(\frac{v}{\sqrt{2}\sigma}\right)} = 1 - e^{\frac{-v^2}{800}}$, so

$F\left(25; 2, \sqrt{2}\right) = 1 - e^{\frac{-625}{800}} = 1 - .458 = .542$

Chapter 4: Continuous Random Variables and Probability Distributions

149.

   a. $F(x) = \lambda e^{-\lambda x}$ and $F(x) = 1 - e^{-\lambda x}$, so $r(x) = \dfrac{\lambda e^{-\lambda x}}{e^{-\lambda x}} = \lambda$, a constant; this is

       consistent with the memoryless property of the exponential distribution.

   b. $r(x) = \left(\dfrac{\alpha}{\beta^{\alpha}}\right) x^{\alpha-1}$; for $\alpha > 1$ this is increasing, while for $\alpha < 1$ it is a decreasing

       function.

   c. $\ln(1 - F(x)) = -\displaystyle\int \alpha\left(1 - \dfrac{x}{\beta}\right)dx = -\alpha\left[x - \dfrac{x^2}{2\beta}\right] \Rightarrow F(x) = 1 - e^{-\alpha\left(x - \frac{x^2}{2\beta}\right)}$,

       $f(x) = \alpha\left(1 - \dfrac{x}{\beta}\right)e^{-\alpha\left(x - \frac{x^2}{2\beta}\right)}$ $0 \le x \le \beta$

151.

   a. $E(g(X)) \approx E[g(\mu) + g'(\mu)(X - \mu)] = E(g(\mu)) + g'(\mu)\cdot E(X - \mu)$, but $E(X) - \mu = 0$ and
       $E(g(\mu)) = g(\mu)$ (since $g(\mu)$ is constant), giving $E(g(X)) \approx g(\mu)$.
       $V(g(X)) \approx V[g(\mu) + g'(\mu)(X - \mu)] = V[g'(\mu)(X - \mu)] = (g'(\mu))^2 \cdot V(X - \mu) = $
       $(g'(\mu))^2 \cdot V(X)$.

   b. $g(I) = \dfrac{v}{I}, g'(I) = \dfrac{-v}{I^2}$, so $E\big(g(I)\big) = \mu_R \approx \dfrac{v}{\mu_I} = \dfrac{v}{20}$

       $V\big(g(I)\big) \approx \left(\dfrac{-v}{\mu_I^2}\right)^2 \cdot V(I), \sigma_{g(I)} \approx \dfrac{v}{20^2} \cdot \sigma_I = \dfrac{v}{800}$

153. For $y > 0$, $F(y) = P(Y \le y) = P\left(\dfrac{2X^2}{\beta^2} \le y\right) = P\left(X^2 \le \dfrac{\beta^2 y}{2}\right) = P\left(X \le \dfrac{\beta\sqrt{y}}{\sqrt{2}}\right)$. Now take

       the cdf of X (Weibull), replace x by $\dfrac{\beta\sqrt{y}}{\sqrt{2}}$, and then differentiate with respect to y to

       obtain the desired result $f_Y(y)$.

Chapter 4: Continuous Random Variables and Probability Distributions

155. When $X \leq q$, gross profits are profit + salvage = $dX + e(q - X)$. But when $X > q$, gross profits are profit – shortage cost = $dq - f \cdot (X - q)$. In any case, there are fixed costs of $c_0 + c_1 q$. If we let Y denote the <u>net</u> profit, then $E[Y] = \int_0^\infty [\text{gross profit}]\, f_X(x)\, dx - [c_0 + c_1 q] = \int_0^q [dx + e(q - x)] f_X(x)\, dx + \int_q^\infty [dq - f(x - q)] f_X(x)\, dx - [c_0 + c_1 q]$. Expand and simplify: $E[Y] = (d - e) \int_0^q x f_X(x)\, dx + eq\, F_X(q) + (dq + fq)[1 - F_X(q)] - f \int_q^\infty x f_X(x)\, dx - [c_0 + c_1 q]$.

Differentiate, using the Fundamental Theorem of Calculus, and then cancel as much as possible: $d/dq\, E[Y] = (d - e)q f_X(q) + e F_X(q) + eq f_X(q) + (d + f)[1 - F_X(q)] + (dq + fq)[-f_X(q)] - [-f\, q f_X(q)] - [0 + c_1] = e F_X(q) + (d + f)[1 - F_X(q)] - c_1$. Whew! The optimal value q^* makes the derivative equal zero, so $e F_X(q^*) + (d + f)[1 - F_X(q^*)] - c_1 = 0$, from which we finally get $F_X(q^*) = (d - c_1 + f)/(d - e + f)$. Notice that the fixed cost c_0 is irrelevant to the optimization. For the values provided, $F_X(q^*) = (35 - 15 + 25)/(35 - 5 + 25) = 45/55 = .8182$.

Chapter 5: Joint Probability Distributions and Random Samples

1.

  a. $P(X = 1, Y = 1) = p(1,1) = .20$

  b. $P(X \le 1 \text{ and } Y \le 1) = p(0,0) + p(0,1) + p(1,0) + p(1,1) = .42$

  c. At least one hose is in use at both islands. $P(X \ne 0 \text{ and } Y \ne 0) = p(1,1) + p(1,2) + p(2,1) + p(2,2) = .70$

  d. By summing row probabilities, $p_x(x) = .16, .34, .50$ for $x = 0, 1, 2$, and by summing column probabilities, $p_y(y) = .24, .38, .38$ for $y = 0, 1, 2$. $P(X \le 1) = p_x(0) + p_x(1) = .50$

  e. $P(0,0) = .10$, but $p_x(0) \cdot p_y(0) = (.16)(.24) = .0384 \ne .10$, so X and Y are not independent.

3.

  a. $p(1,1) = .15$, the entry in the 1st row and 1st column of the joint probability table.

  b. $P(X_1 = X_2) = p(0,0) + p(1,1) + p(2,2) + p(3,3) = .08 + .15 + .10 + .07 = .40$

  c. $A = \{ (x_1, x_2): x_1 \ge 2 + x_2 \} \cup \{ (x_1, x_2): x_2 \ge 2 + x_1 \}$
    $P(A) = p(2,0) + p(3,0) + p(4,0) + p(3,1) + p(4,1) + p(4,2) + p(0,2) + p(0,3) + p(1,3) = .22$

  d. $P(\text{exactly } 4) = p(1,3) + p(2,2) + p(3,1) + p(4,0) = .17$
    $P(\text{at least } 4) = P(\text{exactly } 4) + p(4,1) + p(4,2) + p(4,3) + p(3,2) + p(3,3) + p(2,3) = .46$

5.

  a. $P(X = 3, Y = 3) = P(3 \text{ customers, each with 1 package})$
    $= P(\text{ each has 1 package } | \text{ 3 customers}) \cdot P(3 \text{ customers})$
    $= (.6)^3 \cdot (.25) = .054$

  b. $P(X = 4, Y = 11) = P(\text{total of 11 packages } | \text{ 4 customers}) \cdot P(4 \text{ customers})$
    Given that there are 4 customers, there are 4 different ways to have a total of 11 packages: 3, 3, 3, 2 or 3, 3, 2, 3 or 3, 2, 3 ,3 or 2, 3, 3, 3. Each way has probability $(.1)^3(.3)$, so $p(4, 11) = 4(.1)^3(.3)(.15) = .00018$

Chapter 5: Joint Probability Distributions and Random Samples

7.

    a. $p(1,1) = .030$

    b. $P(X \le 1 \text{ and } Y \le 1 = p(0,0) + p(0,1) + p(1,0) + p(1,1) = .120$

    c. $P(X = 1) = p(1,0) + p(1,1) + p(1,2) = .100$; $P(Y = 1) = p(0,1) + \dots + p(5,1) = .300$

    d. $P(\text{overflow}) = P(X + 3Y > 5) = 1 - P(X + 3Y \le 5) = 1 - P[(X,Y)=(0,0) \text{ or } \dots \text{ or }$
       $(5,0) \text{ or } (0,1) \text{ or } (1,1) \text{ or } (2,1)] = 1 - .620 = .380$

    e. The marginal probabilities for X (row sums from the joint probability table) are
       $p_x(0) = .05$, $p_x(1) = .10$, $p_x(2) = .25$, $p_x(3) = .30$, $p_x(4) = .20$, $p_x(5) = .10$; those
       for Y (column sums) are $p_y(0) = .5$, $p_y(1) = .3$, $p_y(2) = .2$. It is now easily verified
       that for every (x,y), $p(x,y) = p_x(x) \cdot p_y(y)$, so X and Y are independent.

9.

    a. $1 = \displaystyle\int_{-\infty}^{\infty} \int_{-\infty}^{\infty} f(x, y)\,dxdy = \int_{20}^{30} \int_{20}^{30} K(x^2 + y^2)\,dxdy$

       $= K \displaystyle\int_{20}^{30} \int_{20}^{30} x^2\,dydx + K \int_{20}^{30} \int_{20}^{30} y^2\,dxdy = 10K \int_{20}^{30} x^2\,dx + 10K \int_{20}^{30} y^2\,dy$

       $= 20K \cdot \left(\dfrac{19,000}{3} \right) \Rightarrow K = \dfrac{3}{380,000}$

    b. $P(X < 26 \text{ and } Y < 26) = \displaystyle\int_{20}^{26} \int_{20}^{26} K(x^2 + y^2)\,dxdy = 12K \int_{20}^{26} x^2\,dx = $

       $4Kx^3 \Big|_{20}^{26} = 38,304K = .3024$

    c.

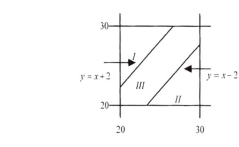

       $P(\,|X - Y| \le 2\,) = \displaystyle\iint_{\substack{region \\ III}} f(x, y)\,dxdy$

94

Chapter 5: Joint Probability Distributions and Random Samples

$$1 - \iint_{I} f(x,y)dxdy - \iint_{II} f(x,y)dxdy$$

$$1 - \int_{20}^{28}\int_{x+2}^{30} f(x,y)dydx - \int_{22}^{30}\int_{20}^{x-2} f(x,y)dydx$$

$$= \text{(after much algebra) } .3593$$

d. $f_x(x) = \int_{-\infty}^{\infty} f(x,y)dy = \int_{20}^{30} K(x^2 + y^2)dy = 10Kx^2 + K\dfrac{y^3}{3}\Big|_{20}^{30}$

$$= 10Kx^2 + .05, \qquad\qquad 20 \le x \le 30$$

e. $f_y(y)$ is obtained by substituting y for x in (d); clearly $f(x,y) \ne f_x(x) \cdot f_y(y)$, so X and Y are not independent.

11.

a. $p(x,y) = \dfrac{e^{-\lambda}\lambda^x}{x!} \cdot \dfrac{e^{-\mu}\mu^y}{y!}$ for x = 0, 1, 2, ...; y = 0, 1, 2, ...

b. $p(0,0) + p(0,1) + p(1,0) = e^{-\lambda-\mu}\left[1 + \lambda + \mu\right]$

c. $P(X+Y=m) = \displaystyle\sum_{k=0}^{m} P(X = k, Y = m-k) = \sum_{k=0}^{m} e^{-\lambda-\mu}\dfrac{\lambda^k}{k!}\dfrac{\mu^{m-k}}{(m-k)!}$

$\dfrac{e^{-(\lambda+\mu)}}{m!}\displaystyle\sum_{k=0}^{m}\binom{m}{k}\lambda^k\mu^{m-k} = \dfrac{e^{-(\lambda+\mu)}(\lambda+\mu)^m}{m!}$, so the total # of errors X+Y also

has a Poisson distribution with parameter $\lambda + \mu$.

13.

a. $f(x,y) = f_x(x) \cdot f_y(y) = \begin{cases} e^{-x-y} & x \ge 0, y \ge 0 \\ 0 & otherwise \end{cases}$

b. $P(X \le 1 \text{ and } Y \le 1) = P(X \le 1) \cdot P(Y \le 1) = (1 - e^{-1})(1 - e^{-1}) = .400$

c. $P(X + Y \le 2) = \displaystyle\int_0^2\int_0^{2-x} e^{-x-y} dydx = \int_0^2 e^{-x}\left[1 - e^{-(2-x)}\right]dx$

$$= \int_0^2 (e^{-x} - e^{-2})dx = 1 - e^{-2} - 2e^{-2} = .594$$

d. $P(X + Y \le 1) = \displaystyle\int_0^1 e^{-x}\left[1 - e^{-(1-x)}\right]dx = 1 - 2e^{-1} = .264$,

so $P(1 \le X + Y \le 2) = P(X + Y \le 2) - P(X + Y \le 1) = .594 - .264 = .330$

Chapter 5: Joint Probability Distributions and Random Samples

15.

  a. $F(y) = P(\ Y \le y\ ) = P\ [(X_1 \le y) \cup ((X_2 \le y) \cap (X_3 \le y))]$
  $= P\ (X_1 \le y) + P[(X_2 \le y) \cap (X_3 \le y)] - P[(X_1 \le y) \cap (X_2 \le y) \cap (X_3 \le y)]$
  $= (1 - e^{-\lambda y}) + (1 - e^{-\lambda y})^2 - (1 - e^{-\lambda y})^3$ for $y \ge 0$

  b. $f(y) = F'(y) = \lambda e^{-\lambda y} + 2(1 - e^{-\lambda y})(\lambda e^{-\lambda y}) - 3(1 - e^{-\lambda y})^2(\lambda e^{-\lambda y})$
  $= 4\lambda e^{-2\lambda y} - 3\lambda e^{-3\lambda y}$ for $y \ge 0$

  $$E(Y) = \int_0^\infty y \cdot (4\lambda e^{-2\lambda y} - 3\lambda e^{-3\lambda y}) dy = 2\left(\frac{1}{2\lambda}\right) - \frac{1}{3\lambda} = \frac{2}{3\lambda}$$

17.

  a. $P((X,Y)$ within a circle of radius $\frac{R}{2}) = P(A) = \iint_A f(x,y)dxdy$

  $$= \frac{1}{\pi R^2} \iint_A dxdy = \frac{area.of.A}{\pi R^2} = \frac{1}{4} = .25$$

  b.

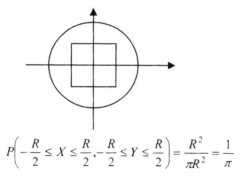

  $$P\left(-\frac{R}{2} \le X \le \frac{R}{2}, -\frac{R}{2} \le Y \le \frac{R}{2}\right) = \frac{R^2}{\pi R^2} = \frac{1}{\pi}$$

  c.

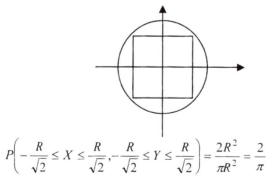

  $$P\left(-\frac{R}{\sqrt{2}} \le X \le \frac{R}{\sqrt{2}}, -\frac{R}{\sqrt{2}} \le Y \le \frac{R}{\sqrt{2}}\right) = \frac{2R^2}{\pi R^2} = \frac{2}{\pi}$$

Chapter 5: Joint Probability Distributions and Random Samples

d. $f_x(x) = \int_{-\infty}^{\infty} f(x,y)dy = \int_{-\sqrt{R^2-x^2}}^{\sqrt{R^2-x^2}} \frac{1}{\pi R^2}dy = \frac{2\sqrt{R^2-x^2}}{\pi R^2}$ for $-R \le x \le R$ and

similarly for $f_Y(y)$. X and Y are not independent since e.g. $f_x(.9R) = f_Y(.9R) > 0$, yet $f(.9R, .9R) = 0$ since $(.9R, .9R)$ is outside the circle of radius R.

19. $E(X_1 - X_2) = \sum_{x_1=0}^{4}\sum_{x_2=0}^{3}(x_1 - x_2)\cdot p(x_1, x_2) =$

$(0-0)(.08) + (0-1)(.07) + \ldots + (4-3)(.06) = .15$
(which also equals $E(X_1) - E(X_2) = 1.70 - 1.55$)

21. $E(XY) = E(X)\cdot E(Y) = L\cdot L = L^2$

23. $E[h(X,Y)] = \int_0^1\int_0^1 |x-y|\cdot f(x,y)dxdy = \int_0^1\int_0^1 |x-y|\cdot 6x^2 ydxdy$

$= \int_0^1\int_0^x (x-y)\cdot 6x^2 ydydx + \int_0^1\int_x^1 (x-y)\cdot 6x^2 ydydx = \frac{1}{6} + \frac{1}{12} = \frac{1}{4}$

25. $\text{Cov}(X,Y) = -\frac{2}{75}$ and $\mu_x = \mu_y = \frac{2}{5}$. $E(X^2) = \int_0^1 x^2 \cdot f_x(x)dx$

$= 12\int_0^1 x^3(1-x^2 dx) = \frac{12}{60} = \frac{1}{5}$, so Var (X) $= \frac{1}{5} - \frac{4}{25} = \frac{1}{25}$

Similarly, Var(Y) $= \frac{1}{25}$, so $\rho_{X,Y} = \frac{-\frac{2}{75}}{\sqrt{\frac{1}{25}}\cdot\sqrt{\frac{1}{25}}} = -\frac{50}{75} = -.667$

27.

a. $E(X) = \int_{20}^{30} xf_x(x)dx = \int_{20}^{30} x[10Kx^2 + .05]dx = 25.329 = E(Y)$

$E(XY) = \int_{20}^{30}\int_{20}^{30} xy\cdot K(x^2 + y^2)dxdy = 641.447$

$\Rightarrow \text{Cov}(X,Y) = 641.447 - (25.329)^2 = -.111$

b. $E(X^2) = \int_{20}^{30} x^2[10Kx^2 + .05]dx = 649.8246 = E(Y^2)$,

so Var (X) = Var(Y) = $649.8246 - (25.329)^2 = 8.2664$

$\Rightarrow \rho = \frac{-.111}{\sqrt{(8.2664)(8.2664)}} = -.0134$

Chapter 5: Joint Probability Distributions and Random Samples

29. Since $E(XY) = E(X) \cdot E(Y)$, $Cov(X,Y) = E(XY) - E(X) \cdot E(Y) = E(X) \cdot E(Y) - E(X) \cdot E(Y) = 0$, and since $Corr(X,Y) = \dfrac{Cov(X,Y)}{\sigma_x \sigma_y}$, then $Corr(X,Y) = 0$

31.

    a. $Cov(aX + b, cY + d) = E[(aX + b)(cY + d)] - E(aX + b) \cdot E(cY + d)$
       $= E[acXY + adX + bcY + bd] - (aE(X) + b)(cE(Y) + d)$
       $= acE(XY) - acE(X)E(Y) = acCov(X,Y)$

    b. $Corr(aX + b, cY + d) =$
       $\dfrac{Cov(aX + b, cY + d)}{\sqrt{Var(aX + b)}\sqrt{Var(cY + d)}} = \dfrac{acCov(X,Y)}{|a| \cdot |c| \sqrt{Var(X) \cdot Var(Y)}}$
       $= Corr(X,Y)$ when a and c have the same signs.

    c. When a and c differ in sign, $Corr(aX + b, cY + d) = -Corr(X,Y)$.

33. $Cov(aX+bY,Z) = E[(aX+bY-(a\mu_X+b\mu_Y))\cdot(Z - \mu_Z)] = E[(aX - a\mu_X)(Z - \mu_Z) + (bY - b\mu_Y)(Z - \mu_Z)] = aE[(X - \mu_X)(Z - \mu_Z)] + bE[(Y - \mu_Y)(Z - \mu_Z)] = a\,Cov(X,Z) + b\,Cov(Y,Z)$.

35. Remember that for any standardized rv Z, $E[Z^2] = V(Z) + [E(Z)]^2 = 1 + 0^2 = 1$.

    a. $E[(Z_Y-\rho Z_X)^2] = E[Z_Y^2] - 2\rho E[Z_Y Z_X] + \rho^2 E[Z_X^2]$. From Exercise 34 and the remark above, this becomes $1 - 2\rho Corr(X,Y) + \rho^2(1) = 1 - 2\rho^2 + \rho^2 = 1 - \rho^2$.

    b. Since $(Z_Y-\rho Z_X)^2 \geq 0$ --- after all, it's a square --- $E[(Z_Y-\rho Z_X)^2] \geq 0$. From part (a), this says $1 - \rho^2 \geq 0$; i.e., $-1 \leq \rho \leq 1$.

    c. From (a), $\rho = 1 \rightarrow E[(Z_Y-Z_X)^2] = 1 - (1)^2 = 0$. Since $(Z_Y-Z_X)^2$ is non-negative, this is only possible if $Z_Y-Z_X = 0$; i.e. $Z_Y = Z_X$. Unraveling the standardized formulas gives $Y = aX+b$ with $a = \sigma_Y/\sigma_X > 0$. Similarly, $\rho = -1 \rightarrow Y = aX+b$ with $a = -\sigma_Y/\sigma_X < 0$.

37.

    a. $f_X(x) = \int_0^x f(x, y)dy = \int_0^x 2dy = 2x,\ 0 < x < 1$.

    b. $f_{Y|X}(y|x) = f(x,y)/f_X(x) = 2/2x = 1/x,\ 0 < y < x$. That is, $Y|X=x$ is Uniform on $(0,x)$. We will use this repeatedly in what follows.

    c. From (b), $P(0<Y<.3|X=.5) = .3/.5 = .6$.

    d. NO, their values are restricted by $y < x$.

    e. From (b), $E(Y|X=x) = (0+x)/2 = x/2$. Yes, $E(Y|X=x)$ is linear in x.

    f. From (b), $V(Y|X=x) = (x - 0)^2/12 = x^2/12$.

Chapter 5: Joint Probability Distributions and Random Samples

39.

a. $f_X(x) = \int_x^\infty f(x,y)dy = \int_x^\infty 2e^{-(x+y)}dy = 2e^{-2x}$, $x > 0$.

b. $f_{Y|X}(y|x) = f(x,y)/f_X(x) = 2e^{-(x-y)}/2e^{-2x} = e^{x-y}$, $x < y$.

c. $P(Y>2 \mid X=1) = \int_2^\infty f_{Y|X}(y\mid 1)dy = \int_2^\infty e^{1-y}dy = e^{-1} = .3679$.

d. NO, since $f_{Y|X}(y|x)$ actually depends on x.

e. $E(Y|X=x) = \int_x^\infty ye^{x-y}dy = e^x \int_x^\infty ye^{-y}dy = e^x(1+x)e^{-x}$ using integration by parts.

   That is, $E(Y|X=x) = 1+x$, which is linear in x.

f. Using integration by parts and proceeding as in (e), $E(Y^2|X=x) = \ldots = x^2 + 2x + 2$. Thus, $V(Y|X=x) = x^2 + 2x + 2 - (1+x)^2 = 1$.

41.

a. $Y|X=x$ is Uniform(0,x). So, $E(Y|X=x) = (0+x)/2 = x/2$, which is indeed linear in x, and $V(Y|X=x) = (x-0)^2/12 = x^2/12$.

b. $f(x,y) = f_X(x)f_{Y|X}(y|x) = 1/(1-0)\, 1/(x-0) = 1/x$ for $0 < y < x < 1$.

c. $f_Y(y) = \int_y^1 f(x,y)dx = \int_y^1 (1/x)dx = \ln(1) - \ln(y) = -\ln(y)$, $0 < y < 1$. [Note: since $0<y<1$, $\ln(y)$ is actually negative, and the PDF is indeed positive.]

43.

a. $p_{y|x}(y|1)$ results from dividing each entry in x = 1 row of the joint probability table by $p_x(1) = $
   $.34$: $P_{y|x}(0|1) = \dfrac{.08}{.34} = .2353$, $P_{y|x}(1|1) = \dfrac{.20}{.34} = .5882$, $P_{y|x}(2|1) = \dfrac{.06}{.34} = .1765$

b. $P_{y|x}(x|2)$ is requested; to obtain this divide each entry in the y = 2 row by $p_x(2) = .50$:

y	0	1	2
$P_{y|x}(y|2)$	.12	.28	.60

c. $P(Y \le 1 \mid x = 2) = P_{y|x}(0|2) + P_{y|x}(1|2) = .12 + .28 = .40$

d. $P_{X|Y}(x|2)$ results from dividing each entry in the y = 2 column by $p_y(2) = .38$:

x	0	1	2
$P_{x|y}(x|2)$	.0526	.1579	.7895

Chapter 5: Joint Probability Distributions and Random Samples

45.

   a. $Y|X=x$ is Uniform$(0,x^2)$. So, $E(Y|X=x) = (0+ x^2)/2 = x^2/2$, which is not linear in x, and $V(Y|X=x) = (x^2 - 0)^2/12 = x^4/12$.

   b. $f(x,y) = f_X(x)f_{Y|X}(y|x) = 1/(1-0) \cdot 1/(x^2 - 0) = 1/x^2$ for $0 < y < x^2 < 1$.

   c. $f_Y(y) = \int f(x, y)dx = \int_{\sqrt{y}} (1/x^2)dx = \dfrac{1}{\sqrt{y}} - 1, 0 < y < 1.$

47.

   a. By considering all 9 possible pairs of numbers David and Peter could select, we find the joint pmf p(x,y) displayed in the table below.

x\y	1	2	3
1	1/9	0	0
2	2/9	1/9	0
3	2/9	2/9	1/9

   b. Add across the rows of (a).

x	1	2	3
p(x)	1/9	3/9	5/9

   c. $p_{Y|X}(y|x) = p(x,y)/p_X(x)$ for each x and y. This gives three conditional distributions, for each of $x = 1,2,3$.

y	1	2	3
p(y|1)	1	0	0

y	1	2	3
p(y|2)	2/3	1/3	0

y	1	2	3
p(y|3)	2/5	2/5	1/5

   d. From (c), $E(Y|X=1) = 1$; $E(Y|X=2) = 1(2/3)+2(1/3) = 4/3$; and $E(Y|X=3) = 1(2/5)+2(2/5)+3(1/5) = 9/5$. This is not a linear progression, so $E(Y|X=x)$ is not linear in x.

   e. From (c), $V(Y|X=1) = 0$, $V(Y|X=2) = 2/9$ by direct computation, and $V(Y|X=3) = 14/25$ by direct computation.

49.

   a. $p_{X|Y}(x|y) = p(x,y)/p_X(x)$ for each x and y. This gives three conditional distributions, for each of $y = 1,2,3$.

x	1	2	3
p(x|1)	1/5	2/5	2/5

x	1	2	3
p(x|2)	0	1/3	2/3

x	1	2	3
p(x|3)	0	0	1

   b. From (a), $E(X|Y=1) = 11/5$; $E(X|Y=2) = 8/3$; and $E(X|Y=3) = 3$. This is not a linear progression, so $E(X|Y=y)$ is not linear in y.

   c. From (a), $V(X|Y=1) = 14/25$ by direct computation, $V(X|Y=2) = 2/9$ by direct computation, and $V(X|Y=3) = 0$.

Chapter 5: Joint Probability Distributions and Random Samples

51.

    a. $\mu = \mu_2 + \rho\sigma_2(x - \mu_1)/\sigma_1 = 30 + (.8)(5)(x - 20)/2 = 2x - 10$.

    b. $\sigma^2 = \sigma_2^2(1 - \rho)^2 = 5^2(1 - .8^2) = 9$.

    c.   From (b), $\sigma = 3$.

    d.   From (a), the mean when x=25 is 40. So, $P(Y>46|X=25) = P(Z > [46-40]/3) = P(Z > 2) = 1 - \Phi(2) = .0228$.

53.

    a.   Since all ten digits are equally likely, $p_X(x) = 1/10$ for $x = 0,1,...,9$. Next, $p_{Y|X}(y|x) = 1/9$ for $y = 0,1,...,9$, $y \neq x$. (That is, any of the 9 remaining digits are equally likely.) Combining, $p(x,y) = p_X(x)\, p_{Y|X}(y|x) = 1/90$ for (x,y) satisfying x,y $= 0,1,...,9$, $y \neq x$.

    b. $E(Y|X=x) = \sum_{y \neq x} y\, p_{Y|X}(y|x). = (1/9) \sum_{y \neq x} y = (1/9) [0 + 1 + ... + 9 - x] = (1/9)(45 - x) = 5 - x/9$. Yes, this is linear in x.

55.   We will repeatedly use that $Y|X \sim \mathrm{Bin}(X,.6)$.

    a. $E(Y|X) = X(.6) = .6X$ and $V(Y|X) = X(.6)(1 - .6) = .24X$.

    b. $E(Y) = E[E(Y|X)] = E[.6X] = .6E[X] = .6(100) = 60$.

    c. $V(Y) = V(E(Y|X)) + E[V(Y|X)] = V(.6X) + E(.24X) = .36V(X) + .24E(X) = .36(100) + .24(100) = 60$. Here, we have used that the mean and variance of a Poisson rv are equal.

57.

    a. $P(50<X<100,20<Y<25) = P(X<100,Y<25) - P(X<50,Y<25) - P(X<100,Y<20) + P(X<50,Y<20) = .3333 - .1274 - .1274 + .0625 = .1410$.

    b.   If X and Y are independent, then $P(50<X<100,20<Y<25) = P(50<X<100)P(20<Y<25) = P(-1<Z<0)P(-1<Z<0) = (.3413)^2 = .1165$. This is smaller than (a). When $\rho > 0$, it's more likely that X < 100 (its mean) coincides with Y < 25 (its mean).

59.   Let $Y = X_1+X_2$ and $W = X_2-X_1$, so $X_1 = (Y-W)/2$ and $X_2 = (Y+W)/2$. We will find their joint distribution, and then their marginal distributions. The Jacobian of the transformation is $\det\begin{bmatrix} 1/2 & -1/2 \\ 1/2 & 1/2 \end{bmatrix} = 1/2$. Graph the triangle $0 \leq x_1 \leq x_2 \leq 1$ and transform this into the (y,w) plane. The result is the triangle bounded by y = 0, w=y, and w=2-y. Therefore, on this triangle, the joint pdf of Y and W is $f(y,w) = 2(x_1 + x_2)|1/2| = y$.

    a.   For $0 \leq y \leq 1$, $f_Y(y) = \int_0^y y\,dw = y^2$; for $1 \leq y \leq 2$, $f_Y(y) = \int_0^{2-y} y\,dw = y(2 - y)$.

    b.   For $0 \leq w \leq 1$, $f_W(w) = \int_w^{2-w} y\,dy = ... = 2(1 - w)$.

Chapter 5: Joint Probability Distributions and Random Samples

61. Solving for the X's gives $X_1 = Y_1$, $X_2 = Y_2/Y_1$, and $X_3 = Y_3/Y_2$. The Jacobian of the

transformation is $\det \begin{bmatrix} 1 & - & - \\ 0 & 1/y_1 & - \\ 0 & 0 & 1/y_2 \end{bmatrix} = 1/y_1 y_2$; the dashes indicate partial

derivatives that don't matter. Thus, the joint pdf of the Y's is $8(y_3)|1/y_1 y_2| = 8y_3/y_1 y_2$ for $0 < y_3 < y_2 < y_1 < 1$. The marginal PDF of Y_3 is

$$\int_{y_3} \int_{y_2} \frac{8y_3}{y_1 y_2} dy_1 dy_2 = \int_{y_3} -\frac{8y_3}{y_2} \ln(y_2) dy_2 = \int_{\ln(y_3)} -8y_3 u du = 4y_3[\ln(y_3)]^2, 0 < y_3 < 1.$$

63. If U is Uniform(0,1), then $Y = -2\ln(U)$ has an exponential distribution with parameter ½ (mean 2); see the section on one-variable transformations in the previous chapter. Likewise, $2\pi U$ is Uniform on $(0, 2\pi)$. Hence, Y_1, and Y_2 described here are precisely the random variables that result in Exercise 62, and the transformations $z_1 = \sqrt{y_1} \cos(y_2)$, $z_2 = \sqrt{y_1} \sin(y_2)$ restore the original independent standard normal random variables in that exercise.

65.

a. $f(x) = 1/10 \rightarrow F(x) = x/10 \rightarrow g_5(y) = 5[y/10]^4[1/10] = 5y^4/10^5$, $0 < y < 10$. Hence,

$E(Y_5) = \int_0^{10} y \cdot 5y^4/10^5 dy = 50/6$, or 8.33 minutes.

b. By the same sort of computation as in (a), $E(Y_1) = 10/6$, and so $E(Y_5 - Y_1) = 50/6 - 10/6 = 40/6$, or 6.67 minutes.

c. The median waiting time is Y_3; its pdf is $g_3(y) = \frac{5!}{2!1!2!}[F(y)]^2 f(y)[1-F(y)]^2 = 30y^2(10-y)^2/10^5$ for $0 < y < 10$. By direct integration, or by symmetry, $E(Y_3) = 5$ minutes.

d. $E(Y_5^2) = \int_0^{10} y^2 \cdot 5y^4/10^5 dy = 500/7$, so $V(Y_5) = 500/7 - (50/6)^2 = 125/63 = 1.984$, from which $\sigma = 1.409$ minutes.

67. The joint pdf of the smallest and largest waiting time is $g_{1.5}(y_1,y_5) = \frac{5!}{0!3!0!}[F(y_5)]^0 f(y_5)[F(y_5)-F(y_1)]^3 f(y_1)[1-F(y_1)]^0 = 20(y_5 - y_1)^3/10^5$, $0 < y_1 < y_5 < 10$.

Hence, $g(y_5|y_1=4) = g_{1.5}(4,y_5)/g_1(4) = 20(y_5 - 4)^3/10^5 \div 5(10 - 4)^4/10^5 = 4(y_5 - 4)^3/6^4$ for $4 < y_5 < 10$. From this conditional density, it's easy to compute $E(Y_5-4|Y_1=4) = 4.8$, and so $E(Y_5|Y_1=4) = 4.8+4 = 8.8$ minutes.

Chapter 5: Joint Probability Distributions and Random Samples

69. Let the original times (in hours after noon) be $X_1, ..., X_n$, which we're assuming are Uniform[0,1]. Their pdf is 1 and cdf is x, so the pdf of the kth order statistic is $g_k(y) = \dfrac{n!}{(k-1)!(n-k)!} y^{k-1}(1-y)^{n-k}$, which we can recognize as the Beta(k,n-k+1) distribution. The expected value, from the Beta formulas, is then

$E(Y_k) = \dfrac{k}{k+n-k+1} = \dfrac{k}{n+1}$. The expected ordered arrival times are evenly spaced throughout the hour, 1/(n+1) hours apart

71. $g_i(y) = \dfrac{n!}{(i-1)!(n-i)!}[y^\theta]^{i-1}[1-y^\theta]^{n-i}[\theta y^{\theta-1}] = \dfrac{n!\theta}{(i-1)!(n-i)!} y^{i\theta-1}[1-y^\theta]^{n-i}$.

Thus, $E(Y_i) = \displaystyle\int_0^1 yg_i(y)dy = \dfrac{n!\theta}{(i-1)!(n-i)!}\int_0^1 y^{i\theta}[1-y^\theta]^{n-i}dy$ = [via $u = y^\theta$]

$\dfrac{n!\theta}{(i-1)!(n-i)!}\displaystyle\int_0^1 u^i[1-u]^{n-i}\dfrac{u^{1/\theta-1}}{\theta}du = \dfrac{n!}{(i-1)!(n-i)!}\int_0^1 u^{i+1/\theta-1}[1-u]^{n-i}du$. The

integral is the "kernel" of the Beta($i+1/\theta,n-i+1$) distribution, and so this entire

expression equals $\dfrac{n!}{(i-1)!(n-i)!}\dfrac{\Gamma(i+1/\theta)\Gamma(n-i+1)}{\Gamma(n+1/\theta+1)} = \dfrac{n!\Gamma(i+1/\theta)}{(i-1)!\Gamma(n+1/\theta+1)}$.

Similarly, $E(Y_i^2) = \dfrac{n!\Gamma(i+2/\theta)}{(i-1)!\Gamma(n+2/\theta+1)}$, and the variance can be found from these.

73. $g_{3,8}(y_3,y_8) = \dfrac{10!}{2!4!2!}[F(y_3)]^2[F(y_8)-F(y_3)]^4[1-F(y_8)]^2 f(y_3)f(y_8) = $

$37800(y_3/5)^2(y_8/5-y_3/5)^4(1-y_8/5)^2(1/5)(1/5) = $

$\dfrac{37800}{5^{10}}y_3^2(y_8-y_3)^4(5-y_8)^2, 0 < y_3 < y_8 < 5$

75. Given a continuous pdf f with cdf F, consider 5 probabilities:

$p_1 = P(X \le y_i) = F(y_i)$; $p_2 = P(y_i < X \le y_i + \Delta_1) \approx f(y_i)\Delta_1$;

$p_3 = P(y_i + \Delta_1 < X \le y_j) = F(y_j) - F(y_i + \Delta_1)$;

$p_4 = P(y_j < X \le y_j + \Delta_2) \approx f(y_j)\Delta_2$; and $p_5 = P(y_j + \Delta_2 < X) = 1 - F(y_j + \Delta_2)$.

Here, it's understood that the Δ's are small and that $y_i < y_i + \Delta_1 < y_j < y_j + \Delta_2$. In a random sample of size n, the chance the ith order statistic falls in the second interval and the jth order statistic falls in the fourth interval ($i<j$) is roughly a multinomial probability with the above probabilities. Specifically, the desired counts for the 5 "cells" are $i-1$, 1, $j-i-1$, 1, and $n-j$. That is,

Chapter 5: Joint Probability Distributions and Random Samples

$$P(y_i < Y_i \le y_i + \Delta_1, y_j < Y_j \le y_j + \Delta_2) \approx \frac{n!}{(i-1)!1!(j-i+1)!1!(n-j)!} \times$$

$F(y_i)^{i-1} f(y_i)\Delta_1 [F(y_j) - F(y_i + \Delta_1)]^{j-i-1} f(y_j)\Delta_2 [1 - F(y_j + \Delta_2)]^{n-j}$. Divide
both sides by Δ_1 and Δ_2, then let the Δ's go to zero. The result is the joint pdf of Y_i and
Y_j: $\dfrac{n!}{(i-1)!(j-i+1)!(n-j)!} F(y_i)^{i-1} [F(y_j) - F(y_i)]^{j-i-1} [1 - F(y_j)]^{n-j} f(y_i) f(y_j)$.

77.

a. The joint pdf of the maximum and minimum is

$$\frac{n!}{0!(n-2)!0!}[F(y_n) - F(y_1)]^{n-2} f(y_1)f(y_n) \text{ for } y_1 < y_n.$$ Let W denote the range

(drop the subscript "2") and let $W_1 = Y_1$; then $Y_1 = W_1$ and $Y_n = W_1 + W$. The
Jacobian of this transformation is clearly 1, and so the joint pdf of W and W_1 is
$n(n-1)[F(w+w_1) - F(w_1)]^{n-2} f(w_1)f(w+w_1)$. To find the marginal pdf of the
range, integrate out W_1:

$$f_W(w) = n(n-1)\int [F(w+w_1) - F(w_1)]^{n-2} f(w_1)f(w+w_1)dw_1.$$

b. In the case of the Uniform(0,1) distribution, $F(x) = x$ and $f(x) = 1$. Also, $0 < y_1 < y_n < 1$ implies $0 < w_1 < w+w_1 < 1$, so that $0 < w_1 < 1 - w$. The integral above
becomes

$$f_W(w) = n(n-1)\int_0^{1-w} [w + w_1 - w_1]^{n-2} dw_1 = n(n-1)w^{n-2}\int_0^{1-w} dw_1 = n(n-1)w^{n-2} =$$

$n(n-1)w^{n-2}(1-w)$ for $0 < w < 1$. That is, the sample range has a Beta($n-1,2$)
distribution.

79.

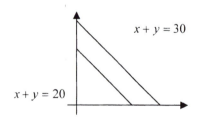

104

Chapter 5: Joint Probability Distributions and Random Samples

a. $1 = \int_{-\infty}^{\infty} \int_{-\infty}^{\infty} f(x, y) dx dy = \int_{0}^{20} \int_{20-x}^{30-x} kxy dy dx + \int_{20}^{30} \int_{0}^{30-x} kxy dy dx$

$= \dfrac{81,250}{3} \cdot k \Rightarrow k = \dfrac{3}{81,250}$

b. $f_X(x) = \begin{cases} \int_{20-x}^{30-x} kxy dy = k(250x - 10x^2) & 0 \le x \le 20 \\ \int_{0}^{30-x} kxy dy = k(450x - 30x^2 + \frac{1}{2}x^3) & 20 \le x \le 30 \end{cases}$

and by symmetry $f_Y(y)$ is obtained by substituting y for x in $f_X(x)$. Since $f_X(25) > 0$, and $f_Y(25) > 0$, but $f(25, 25) = 0$, $f_X(x) \cdot f_Y(y) \ne f(x,y)$ for all x,y so X and Y are not independent.

c. $P(X + Y \le 25) = \int_{0}^{20} \int_{20-x}^{25-x} kxy dy dx + \int_{20}^{25} \int_{0}^{25-x} kxy dy dx$

$= \dfrac{3}{81,250} \cdot \dfrac{230,625}{24} = .355$

d. $E(X + Y) = E(X) + E(Y) = 2 \left\{ \int_{0}^{20} x \cdot k(250x - 10x^2) dx \right.$

$\left. + \int_{20}^{30} x \cdot k(450x - 30x^2 + \frac{1}{2}x^3) dx \right\} = 2k(351,666.67) = 25.969$

e. $E(XY) = \int_{-\infty}^{\infty} \int_{-\infty}^{\infty} xy \cdot f(x, y) dx dy = \int_{0}^{20} \int_{20-x}^{30-x} kx^2 y^2 dy dx$

$+ \int_{20}^{30} \int_{0}^{30-x} kx^2 y^2 dy dx = \dfrac{k}{3} \cdot \dfrac{33,250,000}{3} = 136.4103$, so

Cov(X,Y) = $136.4103 - (12.9845)^2 = -32.19$, and $E(X^2) = E(Y^2) = 204.6154$, so

$\sigma_x^2 = \sigma_y^2 = 204.6154 - (12.9845)^2 = 36.0182$ and

$\rho = \dfrac{-32.19}{36.0182} = -.894$

f. Var $(X + Y) = Var(X) + Var(Y) + 2Cov(X,Y) = 7.66$

Chapter 5: Joint Probability Distributions and Random Samples

81. $E(X+Y-t)^2 = \int_0^1 \int_0^1 (x+y-t)^2 \cdot f(x,y)dxdy$. To find the minimizing value of t, take the derivative with respect to t and equate it to 0:

$$0 = \int_0^1 \int_0^1 2(x+y-t)(-1)f(x,y) = 0 \Rightarrow \int_0^1 \int_0^1 tf(x,y)dxdy = t$$

$$= \int_0^1 \int_0^1 (x+y) \cdot f(x,y)dxdy = E(X+Y) \text{, so the best prediction is the individual's}$$

expected score (= 1.167).

83.

    a.   Suppose *f* and *g* are pdfs and that $f/g \le c$. Then $f(x) \le cg(x)$ for <u>all</u> *x*, and so

        $1 = \int f(x)dx \le \int cg(x)dx = c\int g(x)dx = c(1) = c$. That is, $c \ge 1$.

    b.   P("accept") = $P(U \le f(Y)/cg(Y)) = \int_{-\infty}^{\infty} \int_0^{f(y)/cg(y)} f(u,y)dudy =$

        $\int_{-\infty}^{\infty} \int_0^{f(y)/cg(y)} 1 \cdot g(y)dudy = \int_{-\infty}^{\infty} \frac{f(y)}{cg(y)}g(y)dy = \frac{1}{c}\int_{-\infty}^{\infty} f(y)dy = 1/c$ since $f(y)$ is

        a pdf. As a consequence, the average number of candidates till one is accepted is
        1/[1/c] = c. (You can think of this as a Geometric process, with "success" and
        "accept" being equivalent.) We want 1/c high and c correspondingly low; hence,
        we should try to find the lowest possible *c* which still satisfies $f/g \le c$.

    c.   Let *X* denote the accepted value. Given that *Y* has been "accepted," $X \le x$ iff $Y \le$
        x. That is, $P(X \le x) = P(Y \le x|$ "accept") = $P(Y \le x \cap$ "accept") /P("accept") = P(Y
        $\le x \cap U \le f(Y)/cg(Y))/[1/c]$, from part (b). Calculate the numerator similarly to
        (b): $P(Y \le x \cap U \le f(Y)/cg(Y)) =$

        $\int_{-\infty}^{x} \int_0^{f(y)/cg(y)} f(u,y)dudy = \int_{-\infty}^{x} \int_0^{f(y)/cg(y)} 1 \cdot g(y)dudy =$

        $\int_{-\infty}^{x} \frac{f(y)}{cg(y)}g(y)dy = \frac{1}{c}\int_{-\infty}^{x} f(y)dy = \frac{1}{c}F(x)$. Canceling with the 1/c term in the

        denominator gives $P(X \le x) = F(x)$; i.e., the final result, *X*, really does have *F* as
        its distribution.

    d.   Any distribution with support at least as great as (0,1) will work. The
        Uniform(0,1) distribution is certainly the easiest, and most software has a built-in
        uniform random number generator. We require here that $20x(1-x)^3/1 \le c$ for all *x*
        in (0,1). A graph (or calculus) shows that *f* has a local maximum at *x* = 1/4,
        suggesting the optimal choice of c is $c = f(1/4) = 135/64 = 2.11$.

85.

a.

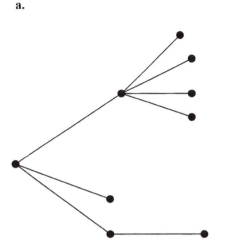

b. By the Law of Total Probability, $A =$

$$\bigcup_{x=0}^{\infty} A \cap \{X_1 = x\} \Rightarrow P(A) = \sum_{x=0}^{\infty} P(A \cap \{X_1 = x\}) =$$

$$\sum_{x=0}^{\infty} P(A \mid X_1 = x)P(X_1 = x) = \sum_{x=0}^{\infty} P(A \mid X_1 = x)p(x). \text{ With } x \text{ members in}$$

generation 1, the process becomes extinct iff these x new, independent branching processes becomes extinct. By definition, the extinction probability for each new branch is $P(A) = p^*$, and independence implies $P(A \mid X_1 = x) = (p^*)^x$. Therefore,

$$p^* = \sum_{x=0}^{\infty} (p^*)^x p(x).$$

c. Check $p^* = 1$: $\sum_{x=0}^{\infty} (1)^x p(x) = \sum_{x=0}^{\infty} p(x) = 1 = p^*$. [We'll drop the * notation from

here forward.] In the first case, we get $p = .3 + .5p + .2p^2$. Solving for p gives $p = 3/2$ and $p = 1$; the smaller value, $p = 1$, is the extinction probability. Why will this model die off with probability 1? Because the expected number of progeny from a single individual is $0(.3)+1(.5)+2(.2) = .9 < 1$. On the other hand, the second case gives $p = .2 + .5p + .3p^2$, whose solutions are $p = 1$ and $p = 2/3$. The extinction probability is the smaller value, $p = 2/3$. Why does this model have positive probability of eternal survival? Because the expected number of progeny from a single individual is $0(.2)+1(.5)+2(.3) = 1.1 > 1$.

Chapter 5: Joint Probability Distributions and Random Samples

87.

a. Use a sort of inclusion-exclusion principle: $P(a \le X \le b, c \le Y \le d) = P(X \le b, Y \le d) - P(X < a, Y \le d) - P(X \le b, Y < c) + P(X < a, Y < c)$. Then, since these variables are continuous, we may write $P(a \le X \le b, c \le Y \le d) = F(b,d) - F(a,d) - F(b,c) + F(a,c)$.

b. In the discrete case, the strict inequalities in (a) must be re-written as follows: $P(a \le X \le b, c \le Y \le d) = P(X \le b, Y \le d) - P(X \le a-1, Y \le d) - P(X \le b, Y \le c-1) + P(X \le a-1, Y \le c-1) = F(b,d) - F(a-1,d) - F(b,c-1) + F(a-1,c-1)$. For the values specified, this becomes $F(10,6) - F(4,6) - F(10,1) + F(4,1)$.

c. Use the cumulative joint cdf table below. At each (x^*, y^*), $F(x^*, y^*)$ is the sum of the probabilities at points (x, y) such that $x \le x^*$ and $y \le y^*$.

$F(x, y)$

			y	
		0	100	250
	100	.20	.30	.50
x	250	.25	.50	1

d. Integrating long-hand and exhausting all possible options for (x,y) pairs, we arrive at the following: $F(x, y) = .6x^2y + .4xy^3$, $0 \le x, y \le 1$; $F(x, y) = 0$, $x \le 0$ or $y \le 0$; $F(x, y) = .6x^2 + .4x$, $0 \le x \le 1, y > 1$; $F(x, y) = .6y + .4y^3$, $x > 1, 0 \le y \le 1$; and, obviously, $F(x, y) = 1$, $x > 1, y > 1$. (Whew!) Thus, from (a), $P(.25 \le x \le .75, .25 \le y \le .75) = F(.75,.75) - F(.25,.75) - F(.75,.25) + F(.25,.25) = \ldots = .23125$. [This only requires the main form of $F(x,y)$; i.e., that for $0 \le x,y \le 1$.]

e. Again, we proceed on a case-by case basis. The results are:
$F(x, y) = 6x^2y^2$, $x + y \le 1, 0 \le x \le 1; 0 \le y \le 1$;
$F(x, y) = 3x^4 - 8x^3 + 6x^2 + 3y^4 - 8y^3 + 6y^2 - 1$, $x + y > 1, x \le 1, y \le 1$;
$F(x, y) = 0$, $x \le 0$; $F(x, y) = 0$, $y \le 0$;
$F(x, y) = 3x^4 - 8x^3 + 6x^2$, $0 \le x \le 1, y > 1$;
$F(x, y) = 3y^4 - 8y^3 + 6y^2$, $0 \le y \le 1, x > 1$; and, $F(x, y) = 1$, $x > 1, y > 1$.

89.

a. For an individual customer, the expected number of packages is $1(.4)+2(.3)+3(.2)+4(.1) = 2$ with a variance of 1 (by direct computation). Given $X=x$, Y is the sum of x independent such customers, so $E(Y|X=x) = x(2) = 2x$ and $V(Y|X=x) = x(1) = x$.

b. $E(Y) = E[E(Y|X)] = E(2X) = 2E(X) = 2(20) = 40$.

c. $V(Y) = V(E(Y|X)) + E(V(Y|X)) = V(2X) + E(X) = 4V(X) + E(X) = 4(20) + 20 = 100$. (Recall that the mean and variance of a Poisson rv are equal.)

91. Let $a = 1/1000$ for notational ease. W is the maximum of the two exponential times, so its pdf is $f_W(w) = 2F_X(w)f_X(w) = 2(1 - e^{-aw})ae^{-aw} = 2ae^{-aw}(1 - e^{-aw})$. From this, $M_W(t)$

$$= E[e^{tW}] = \int_0^\infty e^{tw} 2ae^{-aw}(1 - e^{-aw})dw = 2a\int_0^\infty e^{-(a-t)w}dw - 2a\int_0^\infty e^{-(2a-t)w}dw =$$

$$\frac{2a}{a-t} - \frac{2a}{2a-t} = \frac{2}{(1-1000t)(2-1000t)}. \text{ From this, } E[W] = M_W'(0) = 1500.$$

Chapter 6: Statistics and Sampling Distributions

1.

$P(x_2)$	$P(x_1)$ $x_2 \mid x_1$	.20 25	.50 40	.30 65
.20	25	.04	.10	.06
.50	40	.10	.25	.15
.30	65	.06	.15	.09

a.

$\bar{x}$	25	32.5	40	45	52.5	65
$p(\bar{x})$	.04	.20	.25	.12	.30	.09

$$E(\bar{x}) = (25)(.04) + 32.5(.20) + ... + 65(.09) = 44.5 = \mu$$

b.

s^2	0	112.5	312.5	800
$P(s^2)$	.38	.20	.30	.12

$$E(s^2) = 212.25 = \sigma^2$$

3.

x	0	1	2	3	4	5	6	7	8	9	10
x/n	0	.1	.2	.3	.4	.5	.6	.7	.8	.9	1.0
p(x/n)	.000	.000	.000	.001	.005	.027	.088	.201	.302	.269	.107

X is a binomial random variable with p = .8.

5.

Outcome	1,1	1,2	1,3	1,4	2,1	2,2	2,3	2,4
Probability	.16	.12	.08	.04	.12	.09	.06	.03
$\bar{x}$	1	1.5	2	2.5	1.5	2	2.5	3
r	0	1	2	3	1	0	1	2

Outcome	3,1	3,2	3,3	3,4	4,1	4,2	4,3	4,4
Probability	.08	.06	.04	.02	.04	.03	.02	.01
$\bar{x}$	2	2.5	3	3.5	2.5	3	3.5	4
r	2	1	0	1	3	2	1	2

a.

$\bar{x}$	1	1.5	2	2.5	3	3.5	4
$p(\bar{x})$	.16	.24	.25	.20	.10	.04	.01

b. $P(\bar{x} \le 2.5) = .8$

Chapter 6: Statistics and Sampling Distributions

c.

r	0	1	2	3
p(r)	.30	.40	.22	.08

d. $P(\overline{X} \leq 1.5) = P(1,1,1,1) + P(2,1,1,1) + \ldots + P(1,1,1,2) + P(1,1,2,2) + \ldots +$
   $P(2,2,1,1) + P(3,1,1,1) + \ldots + P(1,1,1,3)$
   $= (.4)^4 + 4(.4)^3(.3) + 6(.4)^2(.3)^2 + 4(.4)^2(.2)^2 = .2400$

7. The MGF of each X_i is $\exp(2(e^t - 1))$, so the MGF of their sum is the product of these 5 MGF's, i.e., $\exp(10(e^t - 1))$. That is to say, ΣX_i is Poisson with parameter 10. The possible values in the sampling distribution of $\overline{X}$ are $\{k/5 : k = 0,1,2,\ldots\}$, and the exact sampling distribution of $\overline{X}$ can be computed by $P(\overline{X} = k/5) = P(\Sigma X_i = k) = e^{-10}10^k/k!$.

9. Use a computer to generate samples of sizes n = 5, 10, 20, and 30 from a Weibull distribution with parameters as given, keeping the number of replications the same, as in problem 43 above. For each sample, calculate the mean. The sampling distribution of $\bar{x}$ for n = 5 appears to be normal, so since larger sample sizes will produce distributions that are closer to normal, the others will also appear normal.

11. $\mu = 12$ cm $\sigma = .04$ cm

  a. n = 16 $E(\overline{X}) = \mu = 12cm$ $\sigma_{\bar{x}} = \dfrac{\sigma_x}{\sqrt{n}} = \dfrac{.04}{4} = .01cm$

  b. n = 64 $E(\overline{X}) = \mu = 12cm$ $\sigma_{\bar{x}} = \dfrac{\sigma_x}{\sqrt{n}} = \dfrac{.04}{8} = .005cm$

  c. $\overline{X}$ is more likely to be within .01 cm of the mean (12 cm) with the second, larger, sample. This is due to the decreased variability of $\overline{X}$ with a larger sample size.

13.

  a. $\mu_{\bar{X}} = \mu = 50$, $\sigma_{\bar{x}} = \dfrac{\sigma_x}{\sqrt{n}} = \dfrac{1}{\sqrt{100}} = .10$

   $P(49.75 \leq \overline{X} \leq 50.25) = P\left(\dfrac{49.75 - 50}{.10} \leq Z \leq \dfrac{50.25 - 50}{.10}\right)$
   $= P(-2.5 \leq Z \leq 2.5) = .9876$

  b. $P(49.75 \leq \overline{X} \leq 50.25) \approx P\left(\dfrac{49.75 - 49.8}{.10} \leq Z \leq \dfrac{50.25 - 49.8}{.10}\right)$
   $= P(-.5 \leq Z \leq 4.5) = .6915$

Chapter 6: Statistics and Sampling Distributions

15.

a. With $\mu = 18\%$, $\sigma = 6\%$, $P(\ 16\% \le \overline{X} \le 19\%) \approx P\left(\dfrac{16-18}{6/\sqrt{40}} \le Z \le \dfrac{19-18}{6/\sqrt{40}}\right) =$

$P(-2.11 \le Z \le 1.05) = \Phi(1.05) - \Phi(-2.11) = .8357$

b. According to the Rule of Thumb, n should be greater than 30 in order to apply the C.L.T., thus using the same procedure for n = 15 as was used for n = 40 would not be appropriate.

17. $X \sim N(10,1)$, n = 4

$\mu_{T_0} = n\mu = (4)(10) = 40$ and $\sigma_{T_0} = \sigma\sqrt{n} = (2)(1) = 2$,

We desire the 95th percentile: $40 + (1.645)(2) = 43.29$

19.

a. $\mu_{\overline{X}} = \mu = 2.65$, $\sigma_{\overline{x}} = \dfrac{\sigma_x}{\sqrt{n}} = \dfrac{.85}{5} = .17$

$P(\overline{X} \le 3.00) = P\left(Z \le \dfrac{3.00 - 2.65}{.17}\right) = P(Z \le 2.06) = .9803$

$P(2.65 \le \overline{X} \le 3.00) == P(\overline{X} \le 3.00) - P(\overline{X} \le 2.65) = .4803$

b. $P(\overline{X} \le 3.00) = P\left(Z \le \dfrac{3.00 - 2.65}{.85/\sqrt{n}}\right) = .99$ implies that $\dfrac{.35}{85/\sqrt{n}} = 2.33$,

from which n = 32.02. Thus n = 33 will suffice.

21.

a. With Y = # of tickets, Y has approximately a normal distribution with

$\mu = \lambda = 50$, $\sigma = \sqrt{\lambda} = 7.071$, so $P(\ 35 \le Y \le 70)$

$\approx P\left(\dfrac{34.5 - 50}{7.071} \le Z \le \dfrac{70.5 - 50}{7.071}\right) = P(\ -2.19 \le Z \le 2.90) = .9838$

b. Here $\mu = 250$, $\sigma^2 = 250$, $\sigma = 15.811$, so $P(\ 225 \le Y \le 275) \approx$

$P\left(\dfrac{224.5 - 250}{15.811} \le Z \le \dfrac{275.5 - 250}{15.811}\right) = P(\ -1.61 \le Z \le 1.61) = .8926$

Chapter 6: Statistics and Sampling Distributions

23. The law of large numbers says that $\overline{X}$ converges to μ; or, equivalently, that $(\overline{X} - \mu)$ converges to zero as $n \to \infty$. The central limit theorem says that if you multiply $(\overline{X} - \mu)$ by the fraction $\dfrac{\sqrt{n}}{\sigma}$, the result is a standard normal random variable as $n \to \infty$.

That is, the inflation factor $\dfrac{\sqrt{n}}{\sigma}$ "balances out" the convergence of $(\overline{X} - \mu)$.

Another way to look at the two theorems is this: roughly, CLT says that (for large n) the sampling distribution of $\overline{X}$ is approximately normal with mean μ and standard deviation $\dfrac{\sigma}{\sqrt{n}}$. As n increases to infinity, this fraction converges to zero, and so the distribution of $\overline{X}$ degenerates into a "normal" distribution with mean μ and standard deviation zero, analogous to saying $\overline{X}$ converges (in some sense) to μ.

25. $P(|Y_n - \theta| \geq \varepsilon) = P(Y_n \geq \theta + \varepsilon) + P(Y_n \leq \theta - \varepsilon) = 0 + P(Y_n \leq \theta - \varepsilon)$, since Y_n obviously can't be greater than θ. Using the CDF of Y_n from Section 5.5, $P(Y_n \leq \theta - \varepsilon) = F_n(\theta - \varepsilon) = \left(\dfrac{\theta - \varepsilon}{\theta}\right)^n$; since $\dfrac{\theta - \varepsilon}{\theta} < 1$, $\left(\dfrac{\theta - \varepsilon}{\theta}\right)^n \to 0$ as $n \to \infty$.

27.

    a. $E(27X_1 + 125X_2 + 512X_3) = 27\,E(X_1) + 125\,E(X_2) + 512\,E(X_3)$
$$= 27(200) + 125(250) + 512(100) = 87{,}850$$
$$V(27X_1 + 125X_2 + 512X_3) = 27^2\,V(X_1) + 125^2\,V(X_2) + 512^2\,V(X_3)$$
$$= 27^2\,(10)^2 + 125^2\,(12)^2 + 512^2\,(8)^2 = 19{,}100{,}116$$

    b. The expected value is still correct, but the variance is not because the covariances now also contribute to the variance.

29. Y is normally distributed with $\mu_Y = \dfrac{1}{2}(\mu_1 + \mu_2) - \dfrac{1}{3}(\mu_3 + \mu_4 + \mu_5) = -1$, and

$$\sigma_Y^2 = \frac{1}{4}\sigma_1^2 + \frac{1}{4}\sigma_2^2 + \frac{1}{9}\sigma_3^2 + \frac{1}{9}\sigma_4^2 + \frac{1}{9}\sigma_5^2 = 3.167, \sigma_Y = 1.7795. \text{ Thus,}$$

$$P(0 \leq Y) = P\left(\frac{0 - (-1)}{1.7795} \leq Z\right) = P(.56 \leq Z) = .2877 \quad \text{and}$$

$$P(-1 \leq Y \leq 1) = P\left(0 \leq Z \leq \frac{2}{1.7795}\right) = P(0 \leq Z \leq 1.12) = .3686$$

Chapter 6: Statistics and Sampling Distributions

31. $E(X_1 + X_2 + X_3) = E(X_1) + E(X_2) + E(X_3) = 15 + 30 + 20 = 65$ min.,

$V(X_1 + X_2 + X_3) = 1^2 + 2^2 + 1.5^2 = 7.25$, $\sigma_{x_1+x_2+x_3} = \sqrt{7.25} = 2.6926$

Thus, $P(X_1 + X_2 + X_3 \leq 60) = P\left(Z \leq \dfrac{60-65}{2.6926}\right) = P(Z \leq -1.86) = .0314$

33. Let $X_1, \ldots, X_5$ denote morning times and $X_6, \ldots, X_{10}$ denote evening times.

    a. $E(X_1 + \ldots + X_{10}) = E(X_1) + \ldots + E(X_{10}) = 5\ E(X_1) + 5\ E(X_6)$
        $= 5(4) + 5(5) = 45$

    b. $\text{Var}(X_1 + \ldots + X_{10}) = \text{Var}(X_1) + \ldots + \text{Var}(X_{10}) = 5\ \text{Var}(X_1) + 5\text{Var}(X_6)$
        $= 5\left[\dfrac{64}{12} + \dfrac{100}{12}\right] = \dfrac{820}{12} = 68.33$

    c. $E(X_1 - X_6) = E(X_1) - E(X_6) = 4 - 5 = -1$
        $\text{Var}(X_1 - X_6) = \text{Var}(X_1) + \text{Var}(X_6) = \dfrac{64}{12} + \dfrac{100}{12} = \dfrac{164}{12} = 13.67$

    d. $E[(X_1 + \ldots + X_5) - (X_6 + \ldots + X_{10})] = 5(4) - 5(5) = -5$
        $\text{Var}[(X_1 + \ldots + X_5) - (X_6 + \ldots + X_{10})]$
        $= \text{Var}(X_1 + \ldots + X_5) + \text{Var}(X_6 + \ldots + X_{10}) = 68.33$

35.

    a. With $M = 5X_1 + 10X_2$, $E(M) = 5(2) + 10(4) = 50$,
        $\text{Var}(M) = 5^2(.5)^2 + 10^2(1)^2 = 106.25$, $\sigma_M = 10.308$.

    b. $P(75 < M) = P\left(\dfrac{75-50}{10.308} < Z\right) = P(2.43 < Z) = .0075$

    c. $M = A_1X_1 + A_2X_2$ with the A_1's and X_1's all independent, so
        $E(M) = E(A_1X_1) + E(A_2X_2) = E(A_1)E(X_1) + E(A_2)E(X_2) = 50$

    d. $\text{Var}(M) = E(M^2) - [E(M)]^2$. Recall that for any r.v. Y,
        $E(Y^2) = \text{Var}(Y) + [E(Y)]^2$. Thus, $E(M^2) = E\left(A_1^2 X_1^2 + 2A_1 X_1 A_2 X_2 + A_2^2 X_2^2\right)$
        $= E\left(A_1^2\right)E\left(X_1^2\right) + 2E(A_1)E(X_1)E(A_2)E(X_2) + E\left(A_2^2\right)E\left(X_2^2\right)$
        (by independence)
        $= (.25 + 25)(.25 + 4) + 2(5)(2)(10)(4) + (.25 + 100)(1 + 16) = 2611.5625$, so
        $\text{Var}(M) = 2611.5625 - (50)^2 = 111.5625$

    e. $E(M) = 50$ still, but now $\text{Cov}(X_1,X_2) = (.5)(.5)(1.0) = .25$, so
        $Var(M) = a_1^2 Var(X_1) + 2a_1 a_2 Cov(X_1, X_2) + a_2^2 Var(X_2)$
        $= 6.25 + 2(5)(10)(.25) + 100 = 131.25$

Chapter 6: Statistics and Sampling Distributions

37.   Let X_1 and X_2 denote the (constant) speeds of the two planes.

    a.   After two hours, the planes have traveled $2X_1$ km. and $2X_2$ km., respectively, so the second will not have caught the first if $2X_1 + 10 > 2X_2$, i.e. if $X_2 - X_1 < 5$. $X_2 - X_1$ has a mean $500 - 520 = -20$, variance $100 + 100 = 200$, and standard deviation 14.14. Thus,

$$P(X_2 - X_1 < 5) = P\left(Z < \frac{5 - (-20)}{14.14}\right) = P(Z < 1.77) = .9616.$$

    b.   After two hours, #1 will be $10 + 2X_1$ km from where #2 started, whereas #2 will be $2X_2$ from where it started. Thus the separation distance will be al most 10 if $|2X_2 - 10 - 2X_1| \le 10$, i.e. $-10 \le 2X_2 - 10 - 2X_1 \le 10$, i.e. $0 \le X_2 - X_1 \le 10$. The corresponding probability is $P(0 \le X_2 - X_1 \le 10) = P(1.41 \le Z \le 2.12) = .9830 - .9207 = .0623$.

39.

    a. $E(Y_i) = .5$, so $E(W) = \sum\limits_{i=1}^{n} i \cdot E(Y_i) = .5 \sum\limits_{i=1}^{n} i = \dfrac{n(n+1)}{4}$

    b. $Var(Y_i) = .25$, so $Var(W) = \sum\limits_{i=1}^{n} i^2 \cdot Var(Y_i) = .25 \sum\limits_{i=1}^{n} i^2 = \dfrac{n(n+1)(2n+1)}{24}$

41.   The total elapsed time between leaving and returning is $T_o = X_1 + X_2 + X_3 + X_4$, with $E(T_o) = 40$, $\sigma_{T_o}^2 = 30$, $\sigma_{T_o} = 5.477$. T_o is normally distributed, and the desired value t is the 99th percentile of the lapsed time distribution added to 10 A.M.: $10:00 + [40 + (5.477)(2.33)] = 10:52.76$.

43.   X is approximately normal with $\mu_1 = (50)(.7) = 35$ and $\sigma_1^2 = (50)(.7)(.3) = 10.5$, as is Y with $\mu_2 = 30$ and $\sigma_2^2 = 12$. Thus $\mu_{X-Y} = 5$ and $\sigma_{X-Y}^2 = 22.5$, so

$$P(-5 \le X - Y \le 5) \approx P\left(\frac{-10}{4.74} \le Z \le \frac{0}{4.74}\right) = P(-2.11 \le Z \le 0) = .4826$$

45.

    a.   See Chapter 4, Exercise 33.

    b.   The mean of the Laplace distribution is 0 by symmetry; its variance is

$$M_X''(0) = \left.\frac{6t^2 + 2}{(1 - t^2)^3}\right|_{t=0} = 2 . \text{ If } X_1, \ldots, X_n \text{ are a random sample from the Laplace}$$

distribution, then the mgf of $T = \sum X_i$ is $1/(1 - t^2)^n$; the standardized mean is

$$Y = \frac{\sum X_i - n(0)}{\sqrt{n(2)}} = \frac{1}{\sqrt{2n}} T, \text{ so } M_Y(t) = M_T(t/\sqrt{2n}) = 1/(1 - t^2/2n)^n.$$

c. With $a = -t^2/2$, the denominator of $M_Y(t)$ has the form $(1 + a/n)^n$, which converges to $\exp(a) = \exp(-t^2/2)$. Since this is the denominator, $M_Y(t)$ converges to $1/\exp(-t^2/2) = \exp(t^2/2)$, the standard normal mgf.

47. If $X \sim \chi_\nu^2$, then X is distributed as the sum of v iid χ_1^2 random variables. By the Central Limit Theorem, X is then approximately normal for large v.

49. Recall from calculus that the maximum of f and $\ln(f)$ occur at the same x-value. If f is the χ_ν^2 pdf, then $\ln(f) = C + (v/2 - 1)\ln(x) - x/2$, where C is a constant. Take the first derivative and set that equal to zero: $0 + (v/2 - 1)/x - 1/2 = 0 \rightarrow x = v - 2$. This is only a valid value for x if $v - 2 > 0$; i.e., $v > 2$.

51. If X_1 and X_2 are independent, then $M_3(t) = M_1(t)M_2(t)$, and so $M_2(t) = M_3(t)/M_1(t)$. Substitute in the given distributions, and $M_2(t) = (1 - 2t)^{-v3/2}/(1 - 2t)^{-v1/2} = (1 - 2t)^{-(v3-v1)/2}$, which is the mgf of the chi-square distribution with $v_3 - v_1$ df. Therefore, by the uniqueness of mgfs, $X_2 \sim \chi_{v_3 - v_1}^2$.

53.

a. From Table A.8, $t_{.005,10} = 3.2$.

b. From Table A.9, $F_{.01,1,10} = 10.04 \approx 3.2^2$. This should be, since $t_{\alpha/2,df}^2 = F_{\alpha,1,df}$.

c. Minitab gives the following:
Inverse Cumulative Distribution Function
```
F distribution with 1 DF in numerator and 10 DF in denominator
P( X <= x )        x
        0.99   10.0443
```

55. E(T) exists iff E(|T|) $< \infty$. But E(|T|) $= \int_{-\infty}^{\infty} \frac{|t|}{\pi(1 + t^2)} dt = 2 \int_0^{\infty} \frac{t}{\pi(1 + t^2)} dt =$

$\frac{1}{\pi} \ln(1 + t^2) \Big|_0^{\infty} = \infty$. That is, E(|T|) diverges, so E(T) does not exist.

Chapter 6: Statistics and Sampling Distributions

57. Let X and Y be independent chi-square rvs with v_1 and v_2 df, respectively.

    a. $E[F_{v_1,v_2}] = E[(X/v_1)\div(Y/v_2)] = v_2/v_1 \, E[X] \, E[1/Y]$. $E[X] = v_1$, and from equation

      6.12, $E[1/Y] = \dfrac{2^{-1}\Gamma(-1+v_2/2)}{\Gamma(v_2/2)} = \dfrac{2^{-1}\Gamma(v_2/2-1)}{(v_2/2-1)\Gamma(v_2/2-1)} = \dfrac{1}{v_2-2}$. Canceling

      gives a final answer of $E[F_{v_1,v_2}] = \dfrac{v_2}{v_2-2}$. This only holds, obviously, if $v_2 > 2$.

    b. By the same process, $E[F_{v_1,v_2}^2] = (v_2/v_1)^2 E[X^2]E[1/Y^2]$. From equation 6.12,

      $E[X^2] = \dfrac{2^2\Gamma(2+v_1/2)}{\Gamma(v_1/2)} = v_1(v_1+2)$ and $E[1/Y^2] = \ldots = \dfrac{1}{(v_2-2)(v_2-4)}$. Put

      together, $E[F_{v_1,v_2}^2] = \dfrac{v_2^2(v_1+2)}{v_1(v_2-2)(v_2-4)}$, and finally $V(F_{v_1,v_2}) =$

      $\dfrac{v_2^2(v_1+2)}{v_1(v_2-2)(v_2-4)} - \left(\dfrac{v_2}{v_2-2}\right)^2 = \ldots = \dfrac{2v_2^2(v_1+v_2-2)}{v_1(v_2-2)^2(v_2-4)}$, for $v_2 > 4$.

59. Let X and Y be independent chi-square rvs with v_1 and v_2 df, respectively. Let $c = F_{p,v_1,v_2}$. Then, by definition, $p = P((X/v_1)\div(Y/v_2) > c) = P((Y/v_2)\div(X/v_1) < 1/c) = 1 - P((Y/v_2)\div(X/v_1) > 1/c) \rightarrow P((Y/v_2)\div(X/v_1) > 1/c) = 1 - p$. Since $(Y/v_2)\div(X/v_1) \sim F_{v_2,v_1}$, $1/c = F_{1-p,v_2,v_1}$. Take reciprocals of both sides to get the desired result.

61.

    a. From Table A.9, $F_{.1,2,4} = 4.32$.

    b. The $F_{2,4}$ pdf is $\dfrac{\Gamma(\frac{1}{2}(2+4))(2/4)^{2/2}x^{(2-2)/2}}{\Gamma(\frac{1}{2}(2))\Gamma(\frac{1}{2}(4))[1+(2/4)x]^{(2+4)/2}} = \dfrac{1}{[1+x/2]^3}$. Let $c = F_{.1,2,4}$;

      then $.1 = \displaystyle\int_c^\infty \dfrac{1}{[1+x/2]^3}\,dx = \dfrac{1}{[1+c/2]^2}$. Solving, $[1+c/2]^2 = 10 \rightarrow c = 2(\sqrt{10} -$

      $1) = 4.3246$.

    c. Minitab gives the following:

      Inverse Cumulative Distribution Function
```
F distribution with 2 DF in numerator and 4 DF in denominator
P( X <= x )           x
       0.9    4.32456
```

63. Use properties of mgfs. If $X \sim$ Gamma(α,β), then $M_X(t) = (1 - \beta t)^{-\alpha}$. Hence, the mgf of cX is $M_X(ct) = (1 - \beta[ct])^{-\alpha} = (1 - [\beta c]t)^{-\alpha}$, which we can identify as the Gamma($\alpha,\beta c$) distribution. In particular, if $X \sim \chi_v^2 =$ Gamma($v/2,2$), then $cX \sim$ Gamma($v/2,2c$).

Chapter 6: Statistics and Sampling Distributions

65.

a. Using the fact that the χ_{50}^2 distribution is approximately normal with mean 50 and variance $2(50) = 100$, $P(\chi_{50}^2 > 70) \approx P(Z > [70-50]/10) = 1 - \Phi(2) = .0228$.

b. Substitute $v = 50$ to get $\chi_{50}^2 \approx 50(1 - 1/225 + Z/15)^3$. Then, $P(\chi_{50}^2 > 70) \approx P(50(1 - 1/225 + Z/15)^3 > 70) = P(Z > 1.847) = 1 - \Phi(1.847) = .03237$. Minitab gives an answer of .032374, suggesting the approximation in (b) is more accurate.

67. The roll-up procedure is not valid for the 75th percentile unless $\sigma_1 = 0$ or $\sigma_2 = 0$ or both σ_1 and $\sigma_2 = 0$, as described below.

Sum of percentiles: $\mu_1 + (Z)\sigma_1 + \mu_2 + (Z)\sigma_2 = \mu_1 + \mu_2 + (Z)(\sigma_1 + \sigma_2)$

Percentile of sums: $\mu_1 + \mu_2 + (Z)\sqrt{\sigma_1^2 + \sigma_2^2}$

These are equal when $Z = 0$ (i.e. for the median) or in the unusual case when $\sigma_1 + \sigma_2 = \sqrt{\sigma_1^2 + \sigma_2^2}$, which happens when $\sigma_1 = 0$ or $\sigma_2 = 0$ or both σ_1 and $\sigma_2 = 0$.

69.

a. Let $X_1, \ldots, X_{12}$ denote the weights for the business-class passengers and $Y_1, \ldots, Y_{50}$ denote the tourist-class weights. Then T = total weight
$= X_1 + \ldots + X_{12} + Y_1 + \ldots + Y_{50} = X + Y$
$E(X) = 12E(X_1) = 12(30) = 360$; $V(X) = 12V(X_1) = 12(36) = 432$.
$E(Y) = 50E(Y_1) = 50(40) = 2000$; $V(Y) = 50V(Y_1) = 50(100) = 5000$.
Thus $E(T) = E(X) + E(Y) = 360 + 2000 = 2360$
And $V(T) = V(X) + V(Y) = 432 + 5000 = 5432$, std dev $= 73.7021$

b. $P(T \leq 2500) = P\left(Z \leq \dfrac{2500 - 2360}{73.7021}\right) = P(Z \leq 1.90) = .9713$

71. $X \sim Bin(200, .45)$ and $Y \sim Bin(300, .6)$. Because both n's are large, both X and Y are approximately normal, so $X + Y$ is approximately normal with mean $(200)(.45) + (300)(.6) = 270$, variance $200(.45)(.55) + 300(.6)(.4) = 121.40$, and standard deviation 11.02. Thus, $P(X + Y \geq 250) = P\left(Z \geq \dfrac{249.5 - 270}{11.02}\right) = P(Z \geq -1.86) = .9686$

73. Ann has 192 oz. The amount which Ann would consume if there were no limit is T_o $= X_1 + \ldots + X_{14}$ where each X_i is normally distributed with $\mu = 13$ and $\sigma = 2$. Thus T_o is normal with $\mu_{T_o} = 182$ and $\sigma_{T_o} = 7.483$, so $P(T_o < 192) = P(Z < 1.34) = .9099$.

Chapter 6: Statistics and Sampling Distributions

75. The student will not be late if $X_1 + X_3 \le X_2$, i.e. if $X_1 - X_2 + X_3 \le 0$. This linear combination has mean -2, variance 4.25, and standard deviation 2.06, so

$$P(X_1 - X_2 + X_3 \le 0) = P\left(Z \le \frac{0-(-2)}{2.06}\right) = P(Z \le .97) = .8340.$$

77.

a. $V(X_1) = V(W + E_1) = \sigma_W^2 + \sigma_E^2 = V(W + E_2) = V(X_2)$ and
$Cov(X_1, X_2) = Cov(W + E_1, W + E_2) = Cov(W, W) + Cov(W, E_2) + Cov(E_1, W) + Cov(E_1, E_2) = Cov(W, W) = V(W) = \sigma_w^2.$

Thus, $\rho = \dfrac{\sigma_W^2}{\sqrt{\sigma_W^2 + \sigma_E^2} \cdot \sqrt{\sigma_W^2 + \sigma_E^2}} = \dfrac{\sigma_W^2}{\sigma_W^2 + \sigma_E^2}$

b. $\rho = \dfrac{1}{1+.0001} = .9999$

79. $E(Y) \doteq h(\mu_1, \mu_2, \mu_3, \mu_4) = 120\left[\frac{1}{10} + \frac{1}{15} + \frac{1}{20}\right] = 26$

The partial derivatives of $h(\mu_1, \mu_2, \mu_3, \mu_4)$ with respect to x_1, x_2, x_3, and x_4 are

$-\dfrac{x_4}{x_1^2}$, $-\dfrac{x_4}{x_2^2}$, $-\dfrac{x_4}{x_3^2}$, and $\dfrac{1}{x_1} + \dfrac{1}{x_2} + \dfrac{1}{x_3}$, respectively. Substituting $x_1 = 10$, $x_2 = 15$, $x_3 = 20$, and $x_4 = 120$ gives -1.2, $-.5333$, $-.3000$, and $.2167$, respectively, so $V(Y) = (1)(-1.2)^2 + (1)(-.5333)^2 + (1.5)(-.3000)^2 + (4.0)(.2167)^2 = 2.6783$, and the approximate sd of y is 1.64.

81. For each pair (x,y), first generate a pair of independent standard normal observations, z_1 and z_2. Exploiting Example 6.15, next create $z_3 = \rho z_1 + \sqrt{1 - \rho^2}\, z_2 = .5z_1 + \sqrt{.75}\, z_2$. Then z_1 and z_3 are bivariate standard normal with correlation .5. Finally, to create the desired (x,y), rescale via $x = 100 + 5z_1$ and $y = 50 + 2z_3$. Now x and y have the desired means and standard deviations, and the rescaling with not affect the correlation (i.e., it's still .5).

Chapter 7: Point Estimation

1.

a. We use the sample mean, $\bar{x}$ to estimate the population mean μ.
$$\hat{\mu} = \bar{x} = \frac{\Sigma x_i}{n} = \frac{3753}{33} = 113.73$$

b. We use the sample median, $\tilde{x} = 113$, to estimate the population median.

c. We use the sample standard deviation, $s = 12.74$.

d. With "success" = IQ greater than 100, x = # of successes = 30, and
$\hat{p} = \frac{30}{33} = .9091$.

e. We use the sample CV, or $\dfrac{s}{\bar{x}} = \dfrac{113.73}{12.74} = 8.927$.

3.

a. We use the sample mean, $\bar{x} = 1.3481$

b. Because we assume normality, the mean = median, so we also use the sample mean, $\bar{x} = 1.3481$. We could also easily use the sample median.

c. We use the 90th percentile of the sample:
$$\hat{\mu} + (1.28)\hat{\sigma} = \bar{x} + 1.28s = 1.3481 + (1.28)(.3385) = 1.7814.$$

d. Since we can assume normality,
$$P(X < 1.5) \approx P\left(Z < \frac{1.5 - \bar{x}}{s}\right) = P\left(Z < \frac{1.5 - 1.3481}{.3385}\right) = P(Z < .45) = .6736$$

e. The estimated standard error of $\bar{x} = \dfrac{\hat{\sigma}}{\sqrt{n}} = \dfrac{s}{\sqrt{n}} = \dfrac{.3385}{\sqrt{16}} = .0846$

5. $n = 5{,}000 \qquad T = 1{,}761{,}300$
$\bar{y} = 374.6 \qquad \bar{x} = 340.6 \qquad \bar{d} = 34.0$
$\hat{\theta}_1 = N\bar{x} = (5{,}000)(340.6) = 1{,}703{,}000$
$\hat{\theta}_2 = T - N\bar{d} = 1{,}761{,}300 - (5{,}000)(34.0) = 1{,}591{,}300$
$\hat{\theta}_3 = T\left(\dfrac{\bar{x}}{\bar{y}}\right) = 1{,}761{,}300\left(\dfrac{340.6}{374.6}\right) = 1{,}601{,}438.281$

Chapter 7: Point Estimation

7.

a. $\hat{\mu} = \bar{x} = \dfrac{\sum x_i}{n} = \dfrac{1206}{10} = 120.6$

b. $\hat{\tau} = 10,000\,\hat{\mu} = 1,206,000$

c. 8 of 10 houses in the sample used at least 100 therms (the "successes"), so
$\hat{p} = \frac{8}{10} = .80$.

d. The ordered sample values are 89, 99, 103, 109, 118, 122, 125, 138, 147, 156,
from which the two middle values are 118 and 122, so
$$\hat{\tilde{\mu}} = \tilde{x} = \dfrac{118+122}{2} = 120.0$$

9.

a. $E(\bar{X}) = \mu = E(X) = \lambda$, so $\bar{X}$ is an unbiased estimator for the Poisson
parameter λ; $\sum x_i = (0)(18) + (1)(37) + ... + (7)(1) = 317$, since n = 150,
$$\hat{\lambda} = \bar{x} = \dfrac{317}{150} = 2.11.$$

b. $\sigma_{\bar{x}} = \dfrac{\sigma}{\sqrt{n}} = \dfrac{\sqrt{\lambda}}{\sqrt{n}}$, so the estimated standard error is $\sqrt{\dfrac{\hat{\lambda}}{n}} = \dfrac{\sqrt{2.11}}{\sqrt{150}} = .119$

11.

a. $E\left(\dfrac{X_1}{n_1} - \dfrac{X_2}{n_2}\right) = \dfrac{1}{n_1}E(X_1) - \dfrac{1}{n_2}E(X_2) = \dfrac{1}{n_1}(n_1 p_1) - \dfrac{1}{n_2}(n_2 p_2) = p_1 - p_2$.

b. $Var\left(\dfrac{X_1}{n_1} - \dfrac{X_2}{n_2}\right) = Var\left(\dfrac{X_1}{n_1}\right) + Var\left(\dfrac{X_2}{n_2}\right) = \left(\dfrac{1}{n_1}\right)^2 Var(X_1) + \left(\dfrac{1}{n_2}\right)^2 Var(X_2)$

$\dfrac{1}{n_1^2}(n_1 p_1 q_1) + \dfrac{1}{n_2^2}(n_2 p_2 q_2) = \dfrac{p_1 q_1}{n_1} + \dfrac{p_2 q_2}{n_2}$, and the standard error is the
square root of this quantity.

Chapter 7: Point Estimation

c. With $\hat{p}_1 = \dfrac{x_1}{n_1}$, $\hat{q}_1 = 1 - \hat{p}_1$, $\hat{p}_2 = \dfrac{x_2}{n_2}$, $\hat{q}_2 = 1 - \hat{p}_2$, the estimated standard

error is $\sqrt{\dfrac{\hat{p}_1 \hat{q}_1}{n_1} + \dfrac{\hat{p}_2 \hat{q}_2}{n_2}}$.

d. $(\hat{p}_1 - \hat{p}_2) = \dfrac{127}{200} - \dfrac{176}{200} = .635 - .880 = -.245$

e. $\sqrt{\dfrac{(.635)(.365)}{200} + \dfrac{(.880)(.120)}{200}} = .041$

13. $E(X) = \int_{-1}^{1} x \cdot \tfrac{1}{2}(1 + \theta x)dx = \dfrac{x^2}{4} + \dfrac{\theta x^3}{6} \Big|_{-1}^{1} = \dfrac{1}{3}\theta$

$E(\overline{X}) = \dfrac{1}{3}\theta \quad \hat{\theta} = 3\overline{X} \Rightarrow E(\hat{\theta}) = E(3\overline{X}) = 3E(\overline{X}) = 3\left(\dfrac{1}{3}\right)\theta = \theta$

15.

a. $E(X^2) = 2\theta$ implies that $E\left(\dfrac{X^2}{2}\right) = \theta$. Consider $\hat{\theta} = \dfrac{\sum X_i^2}{2n}$. Then

$E(\hat{\theta}) = E\left(\dfrac{\sum X_i^2}{2n}\right) = \dfrac{\sum E(X_i^2)}{2n} = \dfrac{\sum 2\theta}{2n} = \dfrac{2n\theta}{2n} = \theta$, implying that $\hat{\theta}$ is

an unbiased estimator for θ.

b. $\sum x_i^2 = 1490.1058$, so $\hat{\theta} = \dfrac{1490.1058}{20} = 74.505$

Chapter 7: Point Estimation

17.

a. $E(\hat{p}) = \sum_{x=0}^{\infty} \frac{r-1}{x+r-1} \cdot \binom{x+r-1}{x} \cdot p^r \cdot (1-p)^x$

$= p \sum_{x=0}^{\infty} \frac{(x+r-2)!}{x!(r-2)!} \cdot p^{r-1} \cdot (1-p)^x = p \sum_{x=0}^{\infty} \binom{x+r-2}{x} p^{r-1}(1-p)^x$

$= p \sum_{x=0}^{\infty} nb(x;r-1,p) = p$.

b. For the given sequence, x = 5, so $\hat{p} = \dfrac{5-1}{5+5-1} = \dfrac{4}{9} = .444$

19.

a. $\lambda = .5p + .15 \Rightarrow 2\lambda = p + .3$, so $p = 2\lambda - .3$ and

$\hat{p} = 2\hat{\lambda} - .3 = 2\left(\dfrac{Y}{n}\right) - .3$; the estimate is $2\left(\dfrac{20}{80}\right) - .3 = .2$.

b. $E(\hat{p}) = E\left(2\hat{\lambda} - .3\right) = 2E\left(\hat{\lambda}\right) - .3 = 2\lambda - .3 = p$, as desired.

c. Here $\lambda = .7p + (.3)(.3)$, so $p = \dfrac{10}{7}\lambda - \dfrac{9}{70}$ and $\hat{p} = \dfrac{10}{7}\left(\dfrac{Y}{n}\right) - \dfrac{9}{70}$.

21.

a. We wish to take the derivative of $\ln\left[\binom{n}{x}p^x(1-p)^{n-x}\right]$, set it equal to zero and

solve for p. $\dfrac{d}{dp}\left[\ln\binom{n}{x} + x\ln(p) + (n-x)\ln(1-p)\right] = \dfrac{x}{p} - \dfrac{n-x}{1-p}$; setting this equal

to zero and solving for p yields $\hat{p} = \dfrac{x}{n}$. For n = 20 and x = 3, $\hat{p} = \dfrac{3}{20} = .15$

b. $E(\hat{p}) = E\left(\dfrac{X}{n}\right) = \dfrac{1}{n}E(X) = \dfrac{1}{n}(np) = p$; thus $\hat{p}$ is an unbiased estimator of p.

c. $(1-.15)^5 = .4437$

Chapter 7: Point Estimation

23.

a. $E(X) = \int_0^1 x(\theta+1)x^\theta dx = \dfrac{\theta+1}{\theta+2} = 1 - \dfrac{1}{\theta+2}$, so the moment estimator $\hat{\theta}$

is the solution to $\overline{X} = 1 - \dfrac{1}{\hat{\theta}+2}$, yielding $\hat{\theta} = \dfrac{1}{1-\overline{X}} - 2$. Since

$\overline{x} = .80, \hat{\theta} = 5 - 2 = 3$.

b. $f(x_1,...,x_n;\theta) = (\theta+1)^n (x_1 x_2 ... x_n)^\theta$, so the log likelihood is

$n\ln(\theta+1) + \theta\sum \ln(x_i)$. Taking $\dfrac{d}{d\theta}$ and equating to 0 yields

$\dfrac{n}{\theta+1} = -\sum \ln(x_i)$, so $\hat{\theta} = -\dfrac{n}{\sum \ln(X_i)} - 1$. Taking $\ln(x_i)$ for each given

x_i yields ultimately $\hat{\theta} = 3.12$.

25. We wish to take the derivative of $\ln\left[\begin{pmatrix} x+r-1 \\ x \end{pmatrix} p^r (1-p)^x \right]$ with respect to p, set

it equal to zero, and solve for p:

$\dfrac{d}{dp}\left[\ln\begin{pmatrix} x+r-1 \\ x \end{pmatrix} + r\ln(p) + x\ln(1-p) \right] = \dfrac{r}{p} - \dfrac{x}{1-p}$. Setting this equal to

zero and solving for p yields $\hat{p} = \dfrac{r}{r+x}$. This is the number of successes over the

total number of trials. The unbiased estimator from Exercise 17 is $\hat{p} = \dfrac{r-1}{r+x-1}$,

which is not the same as the maximum likelihood estimator.

27. By the invariance principle, the mle of $P(X \le 113.4)$ is

$\Phi\left(\dfrac{113.4 - \hat{\mu}}{\hat{\sigma}} \right) = \Phi\left(\dfrac{113.4 - 112.97}{3.91} \right) = \Phi(.11) = .5438$.

Chapter 7: Point Estimation

29.

a. $\left(\frac{x_1}{\theta}\exp\left[-x_1^2/2\theta\right]\right)\cdots\left(\frac{x_n}{\theta}\exp\left[-x_n^2/2\theta\right]\right)=\left(x_1\ldots x_n\right)\dfrac{\exp\left[-\Sigma x_i^2/2\theta\right]}{\theta^n}$.

The natural log of the likelihood function is $\ln\left(x_1\ldots x_n\right)-n\ln(\theta)-\dfrac{\Sigma x_i^2}{2\theta}$.

Taking the derivative wrt θ and equating to 0 gives $-\dfrac{n}{\theta}+\dfrac{\Sigma x_i^2}{2\theta^2}=0$, so

$n\theta=\dfrac{\Sigma x_i^2}{2}$ and $\theta=\dfrac{\Sigma x_i^2}{2n}$. The mle is therefore $\hat{\theta}=\dfrac{\Sigma X_i^2}{2n}$, which is

identical to the unbiased estimator suggested in Exercise 15.

b. For x > 0 the cdf of X if $F(x;\theta)=P(X\le x)$ is equal to $1-\exp\left[\dfrac{-x^2}{2\theta}\right]$.

Equating this to .5 and solving for x gives the median in terms of θ :

$.5=\exp\left[\dfrac{-x^2}{2\theta}\right]$ implies that $\ln(.5)=\dfrac{-x^2}{2\theta}$, so $x=\tilde{\mu}=\sqrt{1.38630\theta}$. The

mle of $\tilde{\mu}$ is therefore $\left(1.38630\hat{\theta}\right)^{\frac{1}{2}}$.

31. The likelihood is $f(y;n,p)=\binom{n}{y}p^y(1-p)^{n-y}$ where

$p=P(X\ge24)=1-\displaystyle\int_0^{24}\lambda e^{-\lambda x}dx=e^{-24\lambda}$. We know $\hat{p}=\dfrac{y}{n}$, so by the

invariance principle $\hat{p}=e^{-24\hat{\lambda}}\Rightarrow\hat{\lambda}=-\dfrac{\ln\hat{p}}{24}=.0120$ for n = 20, y = 15

33. Each $X_i\sim$ Bin(k,p), they're independent, and $X\sim$ Bin(nk,p). The question is whether X is sufficient for p. Let's find out:

$P(\vec{X}=(x_1,\ldots,x_n)\mid X=\Sigma x_i)=\dfrac{P(X_1=x_1,\ldots,X_n=x_n)}{P(X=\Sigma x_i)}=$

124

Chapter 7: Point Estimation

$$\frac{\binom{k}{x_1}p^{x_1}q^{k-x_1}\cdots\binom{k}{x_n}p^{x_n}q^{k-x_n}}{\binom{nk}{\Sigma x_i}p^{\Sigma x_i}q^{nk-\Sigma x_i}}=\frac{\binom{k}{x_1}\cdots\binom{k}{x_n}p^{\Sigma x_i}q^{nk-\Sigma x_i}}{\binom{nk}{\Sigma x_i}p^{\Sigma x_i}q^{nk-\Sigma x_i}}=\frac{\binom{k}{x_1}\cdots\binom{k}{x_n}}{\binom{nk}{\Sigma x_i}}.\text{ This}$$

conditional distribution does not depend on p, so X is sufficient for p. That is, statistician A really doesn't have more information about p than statistician B.

35. Re-write the joint pdf: $f(\vec{x};\alpha,\beta)=\prod_{i=1}^{n}\frac{x_i^{a-1}e^{-x_i/\beta}}{\Gamma(\alpha)\beta^\alpha}=\frac{[\Pi x_i]^{\alpha-1}e^{-\Sigma x_i/\beta}}{[\Gamma(\alpha)\beta^\alpha]^n}$. Let

$g(\Pi x_i,\Sigma x_i;\alpha,\beta)$ be this entire expression (and $h=1$ vacuously). Then, by factorization, $\Pi\,x_i$ and $\sum x_i$ are jointly sufficient for α and β.

37. Let I(A) denote the indicator of an event. Then $f(\vec{x};\theta)=\prod_{i=1}^{n}\frac{1}{2\theta-\theta}I(\theta<x_i<2\theta)=$

$\theta^{-n}I(\theta<x_1,\ldots,x_n<2\theta)=\theta^{-n}I(\theta<x_{\min},x_{\max}<2\theta)$. Set g equal to this entire expression, and we see that $(x_{\min},x_{\max})$ are sufficient for θ.

39. Let Y be the number of items in your sample of 2 that work, so that Y ~ Bin(2,p), and define U = I(Y=1). Then, analogous to example 7.28, E[U] = P(Y=1) = $2pq$. Condition on the sufficient statistic X=x to give the improved estimator U* = E[U|X=x] = P(Y=1|X=x). X-Y is the number of working items in the remainder, so X-Y ~ Bin(n-2,p) and X-Y is independent of Y. Therefore, P(Y=1|X=x) = P(Y=1,X=x)/P(X=x) = P(Y=1,X-Y=x-1)/P(X=x) = P(Y=1)P(X-Y=x-1)/P(X=x) via independence. Plug in the appropriate binomial formulas: U* =

$$2pq\binom{n-2}{x-1}p^{x-1}q^{n-x-1}\Big/\binom{n}{x}p^xq^{n-x}=2\binom{n-2}{x-1}\Big/\binom{n}{x}=2x(n-x)/n(n-1).$$

41.

a. Factor: $f(\mathbf{x};\mu)=C\cdot\exp(-.5\sum(x_i-\mu)^2)=C\cdot\exp(-.5\sum x_i^2)\cdot\exp(\mu\sum x_i-.5n\mu^2)$. Set $h(\mathbf{x})$
$=C\cdot\exp(-.5\sum x_i^2)$ and $g(\sum x_i,\mu)=\exp(\mu\sum x_i-.5n\mu^2)$, and we see that $\sum x_i$ is sufficient for μ. Thus, so is any one-to-one function of $\sum x_i$, like $\bar{x}$.

b. We will use $U^*=E[I(X_1\le c)\,|\,\bar{X}\,]=P(X_1\le c\,|\,\bar{X})$. To find the necessary correlation, use independence and properties of covariance: $\text{Cov}(X_1,\bar{X})=$
$\frac{1}{n}\sum\text{Cov}(X_1,X_i)=\frac{1}{n}[V(X_1)+0+\ldots+0]=\frac{1}{n}[1]=\frac{1}{n}$. Therefore,

$\rho=\dfrac{1/n}{\sqrt{1}\sqrt{1/n}}=\dfrac{1}{\sqrt{n}}$. Then the conditional distribution of x1 given $\bar{X}=\bar{x}$ is

Chapter 7: Point Estimation

normal, with mean value $\mu + \dfrac{1}{\sqrt{n}} \dfrac{\sqrt{1}}{\sqrt{1/n}}(\bar{x} - \mu) = \bar{x}$ and variance $(1)(1 - 1/n) =$

$1 - 1/n$. And so, finally, $U^* = P(X_1 \leq c \mid \overline{X}) = \Phi\left(\dfrac{c - \bar{x}}{\sqrt{1 - 1/n}}\right)$.

43. From before, $\hat{\theta} = Y_n$ has pdf $g(y) = n\theta^{-n}y^{n-1}$. That is, $\hat{\theta}/\theta \sim \text{Beta}(n,1)$.

    a. The mean and variance of the Beta(n,1) distribution are $n/(n+1)$ and $n/(n+1)^2(n+2)$. Let's use these: $E(\tilde{\theta}) = E([(n+1)/n]\hat{\theta}) = [(n+1)/n] \theta n/(n+1) = \theta$. Next, $V(\tilde{\theta}) = V([(n+1)/n]\hat{\theta}) = [(n+1)/n]^2 V(\hat{\theta}) = [(n+1)/n]^2 \theta^2 n/(n+1)^2(n+2) = \theta^2/n(n+2)$.

    b. If we <u>ignore the boundary</u> and say $f(x;\theta) = 1/\theta$, then $\ln[f(x;\theta)] = -\ln(\theta)$ and $S(\theta) = -1/\theta$, from which $I_1(\theta) = E[S^2(\theta)] = 1/\theta^2$. The C-R bound is then $1/nI_1(\theta) = \theta^2/n$.

    c. $V(\tilde{\theta}) < \theta^2/n$, even though $\tilde{\theta}$ is an unbiased estimator of θ. This does not violate the Cramer-Rao theorem, however, because the boundaries of the uniform variable X include θ itself! In these circumstances, Fisher information is not well-defined, and the theorem does not apply. (Note, for example, that if we used the $V(S(\theta))$ version of Fisher information in (b), we'd get zero because $S(\theta)$ is constant.)

45.

    a. In terms of μ and the x's, $\ln[f(\mathbf{x};\mu)] = \# - \sum(x_i - \mu)^2/2\sigma^2 \rightarrow S(\mu) = 0 - \sum 2(x_i - \mu)(-1)/2\sigma^2 = \sum(x_i - \mu)/\sigma^2$. To find the MLE of μ, set this equal to zero, which gives $\sum x_i - n\mu = 0$, or $\mu = (\sum x_i)/n = \bar{x}$.

    b. Since the original X's are normal, we know that $\overline{X}$ is normal with mean μ and variance σ^2/n.

    c. Let's find Fisher information and the C-R bound. For a single observation, our work in (a) shows that $S(\mu) = (X - \mu)/\sigma^2$. So, $I_1(\mu) = E[S^2(\mu)] = E[(X - \mu)^2]/\sigma^4 = \sigma^2/\sigma^4 = 1/\sigma^2$, and the C-R bound is σ^2/n. This is precisely $V(\overline{X})$, so $\overline{X}$ is indeed efficient.

    d. The answer to (b) and the suggested asymptotic distribution agree.

47.

    a. In terms of x and σ, $\ln[f(x;\sigma)] = C - \ln(\sigma) - (x - \mu)^2/2\sigma^2 \rightarrow S(v) = -1/\sigma + (X - \mu)^2/\sigma^3$. So, $I_1(\sigma) = V(S(\sigma)) = V((X - \mu)^2/\sigma^3)$. Use the fact that $(X - \mu)^2/\sigma^2 \sim \chi_1^2$: $V((X - \mu)^2/\sigma^3) = V((X - \mu)^2/\sigma^2)/[\sigma]^2 = 2/\sigma^2$.

    b. The answer in (a) is different from the answer, $1/(2\sigma^4)$, to 46(a), so the information does depend on the parameterization.

Chapter 7: Point Estimation

49. Let x_1 = the time until the first birth, x_2 = the elapsed time between the first and second births, and so on. Then

$$f(x_1,...,x_n;\lambda)=\lambda e^{-\lambda x_1}\cdot(2\lambda)e^{-2\lambda x_2}...(n\lambda)e^{-n\lambda x_n}=n!\,\lambda^n e^{-\lambda\Sigma kx_k}\;.$$ Thus the log

likelihood is $\ln(n!)+n\ln(\lambda)-\lambda\Sigma kx_k$. Taking $\dfrac{d}{d\lambda}$ and equating to 0 yields

$\hat{\lambda}=\dfrac{n}{\displaystyle\sum_{k=1}^{n}kx_k}$. For the given sample, n = 6, x_1 = 25.2, x_2 = 41.7 – 25.2 = 16.5, x_3 = 9.5,

x_4 = 4.3, x_5 = 4.0, x_6 = 2.3; so $\displaystyle\sum_{k=1}^{6}kx_k=(1)(25.2)+(2)(16.5)+...+(6)(2.3)=137.7$

and $\hat{\lambda}=\dfrac{6}{137.7}=.0436$.

51. $MSE(KS^2)=Var(KS^2)+Bias^2(KS^2)$.

$Bias(KS^2)=E(KS^2)-\sigma^2=K\sigma^2-\sigma^2=\sigma^2(K-1)$, and

$$Var(KS^2)=K^2Var(S^2)=K^2\left(E[(S^2)^2]-[E(S^2)]^2\right)=K^2\left(\frac{(n+1)\sigma^4}{n-1}-(\sigma^2)^2\right)$$

$=\dfrac{2K^2\sigma^4}{n-1}\Rightarrow MSE=\left[\dfrac{2K^2}{n-1}+(K-1)^2\right]\sigma^4$. To find the minimizing value of K, take

$\dfrac{d}{dK}$ and equate to 0; the result is $K=\dfrac{n-1}{n+1}$; thus the estimator which minimizes

MSE is neither the unbiased estimator (K = 1) nor the mle $K=\dfrac{n-1}{n}$.

53. With $\displaystyle\sum x=555.86$ and $\displaystyle\sum x^2=15,490$, $s=\sqrt{s^2}=\sqrt{2.1570}=1.4687$.

The $|x_i-\tilde{x}|$'s are, in increasing order, .02, .02, .08, .22, .32, .42, .53, .54, .65, .81,

.91, 1.15, 1.17, 1.30, 1.54, 1.54, 1.71, 2.35, 2.92, 3.50. The median of these values is

$\dfrac{(.81+.91)}{2}=.86$. The estimate based on the resistant estimator is then

$\dfrac{.86}{.6745}=1.275$. This estimate is in reasonably close agreement with s.

Chapter 7: Point Estimation

55.

a. The likelihood is

$$\prod_{i=1}^{n} \frac{1}{\sqrt{2\pi\sigma^2}} e^{-\frac{(x_i-\mu_i)}{2\sigma^2}} \cdot \frac{1}{\sqrt{2\pi\sigma^2}} e^{-\frac{(y_i-\mu_i)}{2\sigma^2}} = \frac{1}{\left(2\pi\sigma^2\right)^n} e^{-\left(\frac{\Sigma(x_i-\mu_i)^2+\Sigma(y_i-\mu_i)^2}{2\sigma^2}\right)}.$$ The log

likelihood is thus $-n\ln\left(2\pi\sigma^2\right) - \frac{\left(\Sigma(x_i-\mu_i)^2+\Sigma(y_i-\mu_i)^2\right)}{2\sigma^2}$. Taking $\dfrac{d}{d\mu_i}$ and equating

to zero gives $\hat{\mu}_i = \dfrac{x_i+y_i}{2}$. Substituting these estimates of the $\hat{\mu}_i$'s into the

log likelihood gives

$$-n\ln\left(2\pi\sigma^2\right) - \frac{1}{2\sigma^2}\left(\Sigma\left(x_i - \frac{x_i+y_i}{2}\right)^2 + \Sigma\left(y_i - \frac{x_i+y_i}{2}\right)^2\right)$$

$$= -n\ln\left(2\pi\sigma^2\right) - \frac{1}{2\sigma^2}\left(\frac{1}{2}\Sigma(x_i - y_i)^2\right).$$ Now taking $\dfrac{d}{d\sigma^2}$, equating to zero,

and solving for σ^2 gives the desired result.

b. $E(\hat{\sigma}) = \dfrac{1}{4n}E\left(\Sigma(X_i - Y_i)^2\right) = \dfrac{1}{4n}\cdot\Sigma E(X_i - Y_i)^2$, but

$E(X_i - Y_i)^2 = V(X_i - Y_i) + \left[E(X_i - Y_i)\right]^2 = 2\sigma^2 + 0 = 2\sigma^2$. Thus

$E(\hat{\sigma}^2) = \dfrac{1}{4n}\Sigma\left(2\sigma^2\right) = \dfrac{1}{4n}2n\sigma^2 = \dfrac{\sigma^2}{2}$, so the mle is definitely not

unbiased; the expected value of the estimator is only half the value of what is
being estimated!

57. Given $X = k$, the investigator must have studied $k+r$ people in total. Let **x** be any
sequence of $k+r$ 0's and 1's with exactly k 0's and r 1's. Then $P((X_1,...,X_{k+r}) = \mathbf{x} \mid X = k) = P((X_1,...,X_{k+r}) = \mathbf{x})/P(X = k) = P((X_1,...,X_{k+r}) = \mathbf{x})/\binom{k+r-1}{r-1}p^r q^k$. But, by
independence, the numerator is just the product of exactly k q's and exactly r p's,
canceling those terms with the denominator; i.e., $P((X_1,...,X_{k+r}) = \mathbf{x} \mid X = k) = p^r q^k/\binom{k+r-1}{r-1}p^r q^k = 1/\binom{k+r-1}{r-1}$, which does not depend on p. Therefore, X is sufficient for
p; knowing the order in which allergy and non-allergy sufferers arrive does not help
estimate p.

Chapter 7: Point Estimation

59. Be careful here: "*s*" refers to the MLE of σ and <u>not</u> the ordinary sample standard deviation! In other words, use $n = 3$ rather than $n-1 = 2$ in your denominator. With the information provided, $c = 150$, $\bar{x} = 150.40$, <u>$s = 3.06$</u>, $k = \sqrt{3/2}$, $w = -.1307$, and $kw = -.16$. Hence, the MVUE for θ is $P(T < -.16(1)/\sqrt{1 - (-.16)^2}) = P(T < -.1621)$, where $T \sim t_1$. Minitab gives .448 for this probability.

In contrast, we may also write $\theta = P(X \le c) = \Phi((c - \mu)/\sigma)$, from which, by the invariance principle, the MLE of θ is $\Phi((c - \bar{x})/s) = \Phi(w) = \Phi(-.16) = .4364$.

61.
$$e^{-2\lambda} = E[\delta(X)] = \sum \delta(x) \frac{e^{-\lambda}\lambda^x}{x!} \Rightarrow e^{-\lambda} = \sum \frac{\delta(x)\lambda^x}{x!} \Rightarrow \sum \frac{(-\lambda)^x}{x!} = \sum \frac{\delta(x)\lambda^x}{x!}.$$

From the uniqueness of Taylor series, these can only be equal if $\delta(X) = (-1)^X$. While unbiased, this estimator is ridiculous: if X happens to be even, we estimate the probability θ to be 1 (no matter whether $X = 0$ or $X = 200$). If X happens to be odd, we estimate θ to be -1!!

63.

a. The points do not fall <u>perfectly</u> on a straight line through the origin, but they come very close to fitting the line $y = 30x$.

b. The joint pdf here is $(2\pi\sigma^2)^{-n/2} \exp\left[-\frac{1}{2\sigma^2}\sum(y_i - \beta x_i)^2\right]$, and so the log-likelihood function is $-\frac{n}{2}\ln(2\pi\sigma^2) - \frac{1}{2\sigma^2}\sum(y_i - \beta x_i)^2 =$

$C - n\ln(\sigma) - \dfrac{\sum(y_i - \beta x_i)^2}{2\sigma^2}$. First, differentiate with respect to β and solve:

$\dfrac{\sum 2(y_i - \beta x_i)(-\beta x_i)}{2\sigma^2} = 0 \rightarrow \hat{\beta} = \dfrac{\Sigma x_i y_i}{\Sigma x_i^2}$. Next, differentiate with respect to σ

and solve: $-\dfrac{n}{\sigma} + \dfrac{\sum(y_i - \hat{\beta}x_i)^2}{\sigma^3} \rightarrow \hat{\sigma}^2 = \dfrac{\sum(y_i - \hat{\beta}x_i)^2}{n}$. For the data

provided, $\hat{\beta} = 30.040$, the estimated minutes per item, and

$\hat{\sigma}^2 = \dfrac{1}{n}\sum(y_i - \hat{\beta}x_i)^2 = 16.912$. When $x = 25$, we would predict $y = 25\hat{\beta} = 571$.

Chapter 8: Statistical Intervals Based on a Single Sample

1.

    a. $z_{\alpha/2} = 2.81$ implies that $\alpha/2 = 1 - \Phi(2.81) = .0025$, so $\alpha = .005$ and the confidence level is $100(1 - \alpha)\% = 99.5\%$.

    b. $z_{\alpha/2} = 1.44$ for $\alpha = 2[1 - \Phi(1.44)] = .15$, and $100(1 - \alpha)\% = 85\%$.

    c. 99.7% implies that $\alpha = .003$, $\alpha/2 = .0015$, and $z_{.0015} = 2.96$. (Look for cumulative area .9985 in the main body of table A.3, the Z table.)

    d. 75% implies $\alpha = .25$, $\alpha/2 = .125$, and $z_{.125} = 1.15$.

3.

    a. A 90% confidence interval will be narrower. The z critical value for a 90% confidence level is 1.645, smaller than the z of 1.96 for the 95% confidence level, thus producing a narrower interval.

    b. Not a correct statement. Once and interval has been created from a sample, the mean μ is either enclosed by it, or not. We have 95% confidence in the general procedure, under repeated and independent sampling.

    c. Not a correct statement. The interval is an estimate for the population mean, not a boundary for population values.

    d. Not a correct statement. In theory, if the process were repeated an infinite number of times, 95% of the intervals would contain the population mean μ. We *expect* 95 out of 100 intervals will contain μ, but we don't know this to be true.

5.

    a. $4.85 \pm \dfrac{(1.96)(.75)}{\sqrt{20}} = 4.85 \pm .33 = (4.52, 5.18)$.

    b. $z_{\alpha/2} = z_{.02/2} = z_{.01} = 2.33$, so the interval is $4.56 \pm \dfrac{(2.33)(.75)}{\sqrt{16}} = (4.12, 5.00)$.

    c. $n = \left[\dfrac{2(1.96)(.75)}{.40}\right]^2 = 54.02$, so n = 55.

    d. $w = 2(.2) = .4$, so $n = \left[\dfrac{2(2.58)(.75)}{.4}\right]^2 = 93.61$, so n = 94.

Chapter 8: Statistical Intervals Based on a Single Sample

7. If $L = 2z_{\alpha/2} \dfrac{\sigma}{\sqrt{n}}$ and we increase the sample size by a factor of 4, the new length is

$L' = 2z_{\alpha/2} \dfrac{\sigma}{\sqrt{4n}} = \left[2z_{\alpha/2} \dfrac{\sigma}{\sqrt{n}}\right]\left(\dfrac{1}{2}\right) = \dfrac{L}{2}$. Thus halving the length requires n to be

increased fourfold. If $n' = 25n$, then $L' = \dfrac{L}{5}$, so the length is decreased by a factor
of 5.

9.

a. $\left(\bar{x} - 1.645\dfrac{\sigma}{\sqrt{n}}, \infty\right)$. From 5a, $\bar{x} = 4.85$, $\sigma = .75$, n = 20;

$4.85 - 1.645\dfrac{.75}{\sqrt{20}} = 4.5741$, so the interval is $(4.5741, \infty)$.

b. $\left(\bar{x} - z_{\alpha}\dfrac{\sigma}{\sqrt{n}}, \infty\right)$

c. $\left(-\infty, \bar{x} + z_{\alpha}\dfrac{\sigma}{\sqrt{n}}\right)$; From 4a, $\bar{x} = 58.3$, $\sigma = 3.0$, n = 25;

$58.3 + 2.33\dfrac{3}{\sqrt{25}} = (-\infty, 59.70)$

11. Y is a binomial r.v. with n = 1000 and p = .95, so E(Y) = np = 950, the expected number of intervals that capture μ, and $\sigma_Y = \sqrt{npq} = 6.892$. Using the normal approximation to the binomial distribution, P(940 ≤ Y ≤ 960) = P(939.5 ≤ Y$_{normal}$ ≤ 960.5) = P(-1.52 ≤ Z ≤ 1.52) = .9357 - .0643 = .8714.

13.

a. $\bar{x} \pm z_{.025}\dfrac{s}{\sqrt{n}} = 1.028 \pm 1.96\dfrac{.163}{\sqrt{69}} = 1.028 \pm .038 = (.990, 1.066)$

b. $w = .05 = \dfrac{2(1.96)(.16)}{\sqrt{n}} \Rightarrow \sqrt{n} = \dfrac{2(1.96)(.16)}{.05} = 12.544 \Rightarrow n = (12.544)^2 \approx 158$

Chapter 8: Statistical Intervals Based on a Single Sample

15.

a. $z_\alpha = .84$, and $\Phi(.84) = .7995 \approx .80$, so the confidence level is 80%.

b. $z_\alpha = 2.05$, and $\Phi(2.05) = .9798 \approx .98$, so the confidence level is 98%.

c. $z_\alpha = .67$, and $\Phi(.67) = .7486 \approx .75$, so the confidence level is 75%.

17. A 95% lower confidence bound for μ is $\bar{x} - z_{.05} \dfrac{s}{\sqrt{n}} = 2.08 - 1.645 \dfrac{7.88}{\sqrt{41}} = 0.056$.

We're 95% confident that $\mu \geq 0.056$; this suggests that the mean population stress change is positive, though potentially by a very small amount. (Notice that if we increased the confidence level to 99%, the lower bound would be negative and we could no longer say with confidence that μ is positive.)

19. $\hat{p} = \dfrac{201}{356} = .5646$; we calculate a 95% confidence interval for the proportion of all dies that pass the probe:

$$\frac{.5646 + \dfrac{(1.96)^2}{2(356)} \pm 1.96 \sqrt{\dfrac{(.5646)(.4354)}{356} + \dfrac{(1.96)^2}{4(356)^2}}}{1 + \dfrac{(1.96)^2}{356}} = \frac{.5700 \pm .0518}{1.01079} = (.513, .615)$$

21. $\hat{p} = \dfrac{133}{539} = .2468$; the 95% lower confidence bound is:

$$\frac{.2468 + \dfrac{(1.645)^2}{2(539)} - 1.645 \sqrt{\dfrac{(.2468)(.7532)}{539} + \dfrac{(1.645)^2}{4(539)^2}}}{1 + \dfrac{(1.645)^2}{539}} = \frac{.2493 - .0307}{1.005} = .218$$

Chapter 8: Statistical Intervals Based on a Single Sample

23.

a. $\hat{p} = \dfrac{24}{37} = .6486$; The 99% confidence interval for p is

$$\frac{.6486 + \dfrac{(2.58)^2}{2(37)} \pm 2.58 \sqrt{\dfrac{(.6486)(.3514)}{37} + \dfrac{(2.58)^2}{4(37)^2}}}{1 + \dfrac{(2.58)^2}{37}} = \frac{.7386 \pm .2216}{1.1799} = (.438, .814)$$

b. $n = \dfrac{2(2.58)^2(.25) - (2.58)^2(.01) \pm \sqrt{4(2.58)^4(.25)(.25 - .01) + .01(2.58)^4}}{.01}$

$= \dfrac{3.261636 \pm 3.3282}{.01} \approx 659$

25.

a. $n = \dfrac{2(1.96)^2(.25) - (1.96)^2(.01) \pm \sqrt{4(1.96)^4(.25)(.25 - .01) + .01(1.96)^4}}{.01} \approx 381$

b. $n = \dfrac{2(1.96)^2\left(\frac{1}{3} \cdot \frac{2}{3}\right) - (1.96)^2(.01) \pm \sqrt{4(1.96)^4\left(\frac{1}{3} \cdot \frac{2}{3}\right)\left(\frac{1}{3} \cdot \frac{2}{3} - .01\right) + .01(1.96)^4}}{.01} \approx 339$

27. Note that the midpoint of the new interval is $\dfrac{x + \dfrac{z^2}{2}}{n + z^2}$, which is roughly $\dfrac{x + 2}{n + 4}$ with a confidence level of 95% and approximating $1.96 \approx 2$. The variance of this quantity is $\dfrac{np(1 - p)}{\left(n + z^2\right)^2}$, or roughly $\dfrac{p(1 - p)}{n + 4}$. Now replacing p with $\dfrac{x + 2}{n + 4}$, we

have $\left(\dfrac{x + 2}{n + 4}\right) \pm z_{\alpha/2} \sqrt{\dfrac{\left(\dfrac{x + 2}{n + 4}\right)\left(1 - \dfrac{x + 2}{n + 4}\right)}{n + 4}}$; For clarity, let $x^* = x + 2$ and

$n^* = n + 4$, then $\hat{p}^* = \dfrac{x^*}{n^*}$ and the formula reduces to $\hat{p}^* \pm z_{\alpha/2} \sqrt{\dfrac{\hat{p}^* \hat{q}^*}{n^*}}$, the

desired conclusion. For further discussion, see the Agresti article.

Chapter 8: Statistical Intervals Based on a Single Sample

29.

    a. 1.341

    b. 1.753

    c. 1.708

    d. 1.684

    e. 2.704

31.

    a. $t_{.025,10} = 2.228$

    b. $t_{.025,15} = 2.131$

    c. $t_{.005,15} = 2.947$

    d. $t_{.005,4} = 4.604$

    e. $t_{.01,24} = 2.492$

    f. $t_{.005,37} \approx 2.712$

33.

    a. A normal probability plot of these 10 values is quite linear.

    b. For these data, $\bar{x} = 38.267$ and $s = 0.2646$. So, a 95% CI for μ is
$$\bar{x} \pm t_{.005,9}\frac{s}{\sqrt{n}} = 38.267 \pm 2.262\frac{0.2646}{\sqrt{10}} = (38.06, 38.47).$$

    c. Simply convert the endpoints into Fahrenheit using F = 1.8C+32: (100.5,101.2). It does appear that guinea pigs are warmer, on average, than humans.

35. $n = 14$, $\bar{x} = 8.48$, $s = .79$; $t_{.05,13} = 1.771$

    a. A 95% lower confidence bound: $8.48 - 1.771\left(\frac{.79}{\sqrt{14}}\right) = 8.48 - .37 = 8.11$. With 95% confidence, the value of the true mean proportional limit stress of all such joints is greater than 8.11 MPa. We must assume that the sample observations were taken from a normally distributed population.

    b. A 95% lower prediction bound: $8.48 - 1.771(.79)\sqrt{1 + \frac{1}{14}} = 8.48 - 1.45 = 7.03$. If this bound is calculated for sample after sample, in the long run 95% of these bounds will provide a lower bound for the corresponding future values of the proportional limit stress of a single joint of this type.

Chapter 8: Statistical Intervals Based on a Single Sample

37. $n = 26$, $\bar{x} = 370.69$, $s = 24.36$; $t_{.05,25} = 1.708$

    a. A 95% upper confidence bound:

$$370.69 + (1.708)\left(\frac{24.36}{\sqrt{26}}\right) = 370.69 + 8.16 = 378.85$$

    b. A 95% upper prediction bound:

$$370.69 + 1.708(24.36)\sqrt{1 + \frac{1}{26}} = 370.69 + 42.45 = 413.14$$

    c. Following a similar argument as in the text, we need to find the variance of
$\bar{X} - X_{new}$: $V(\bar{X} - X_{new}) = V(\bar{X}) + V(X_{new}) = V(\bar{X}) + V\left(\frac{1}{2}(X_{27} + X_{28})\right)$

$$= V(\bar{X}) + V\left(\tfrac{1}{2} X_{27}\right) + V\left(\tfrac{1}{2} X_{28}\right) = V(\bar{X}) + \tfrac{1}{4}V(X_{27}) + \tfrac{1}{4}V(X_{28})$$

$$= \frac{\sigma^2}{n} + \frac{1}{4}\sigma^2 + \frac{1}{4}\sigma^2 = \sigma^2\left(\frac{1}{2} + \frac{1}{n}\right). \text{ We eventually arrive at } T = \frac{\bar{X} - X_{new}}{s\sqrt{\frac{1}{2} + \frac{1}{n}}} \sim t$$

       distribution with $n - 1$ d.f., so the new prediction interval is

$$\bar{x} \pm t_{\alpha/2,n-1} \cdot s\sqrt{\tfrac{1}{2} + \tfrac{1}{n}}. \text{ For this situation, we have}$$

$$370.69 \pm 1.708(24.36)\sqrt{\frac{1}{2} + \frac{1}{26}} = 370.69 \pm 30.53 = (340.16, 401.22)$$

39. $n = 25$, $\bar{x} = .0635$, $s = .0065$

    a. 95% P.I. :

$$.0635 \pm 2.064(.0065)\sqrt{1 + \tfrac{1}{25}} = .0635 \pm .0137 \Rightarrow (.0498, .0772).$$

    b. 99% Tolerance Interval, with $k = 95$, critical value 2.972 (table A.6):

$$.0635 \pm 2.972(.0065) \Rightarrow (.0442, .0828).$$

41.   For this data, $\bar{x} = 174.37$, $s = 19.89$, $n = 63$.

    a. A 95% C.I. : $174.37 \pm t_{.025,62}(19.89)/\sqrt{63} = (169.36, 179.37)$.

    b. A 95% P.I. : $174.37 \pm t_{.025,62}(19.89)\sqrt{1 + \tfrac{1}{63}} = (134.30, 214.43)$, which includes
152.

    c. The second interval is much wider, because it allows for the variability of a
single observation.

    d. The normal probability plot gives no reason to doubt normality. This is
especially important for part (b), but the large sample size implies that normality
is not so critical for (a).

Chapter 8: Statistical Intervals Based on a Single Sample

43.   Let $Z = \dfrac{\overline{X} - X_{n+1}}{\sqrt{\sigma^2 \left(1 + \frac{1}{n}\right)}}$, the standard normal rv suggested in the text. The sample

variance S^2 based on the first n observations is independent of Z: S^2 and $\overline{X}$ are independent for normal random samples, and X_{n+1} is independent of the sample on which S^2 is based. Also, we know that $Y = (n\text{-}1)S^2/\sigma^2 \sim \chi^2_{n-1}$; therefore, the rv

$\dfrac{Z}{\sqrt{Y/(n-1)}}$ has a t distribution with n-1 df. Cancel the common σ^2 terms, and we

find that $\dfrac{Z}{\sqrt{Y/(n-1)}} = \dfrac{\overline{X} - X_{n+1}}{\sqrt{S^2\left(1 + \frac{1}{n}\right)}} = T$, as desired.

45.

   a. $\chi^2_{.05,10} = 18.307$

   b. $\chi^2_{.95,10} = 3.940$

   c.   Since $10.987 = \chi^2_{.975,22}$ and

       $36.78 = \chi^2_{.025,22}$, $P\left(\chi^2_{.975,22} \leq \chi^2 \leq \chi^2_{.025,22}\right) = .95$.

   d.   Since $14.611 = \chi^2_{.95,25}$ and $37.652 = \chi^2_{.05,25}$, $P(\chi^2 < 14.611$ or $\chi^2 > 37.652) =$
       $1 - P(\chi^2 > 14.611) + P(\chi^2 > 37.652) = (1 - .95) + .05 = .10$.

47.   For these data, $\overline{x} = 69.4$, $s = 3.3$, $n = 12$.
   a.   A normal probability plot verifies that the data is plausibly normal.
   b.   At $12 - 1 = 11$df, the chi-square critical values are 3.816 and 21.920. Hence, a

       95% CI for σ is $\left(\sqrt{\dfrac{11}{21.920}}(3.3), \sqrt{\dfrac{11}{3.816}}(3.3)\right) = (1.65, 5.60)$. We are 95%

       confident the standard deviation for performance times of Beethoven's 9th for all
       conductors is between 1.65 minutes and 5.60 minutes.
   c.   False, though we didn't need the CI to confirm this. There's obviously variability
       in the performance times of Beethoven's 9th.

49. For this data, $\bar{x} = 9.955$, $s = 4.603$, $n = 22$.

    a. $t_{.025,21} = 2.080$, so a 95% CI for μ is $9.955 \pm 2.080(4.603)/\sqrt{22} = (7.91, 12.00)$.

    b. The probability plot below shows gross deviation from a straight line, suggesting the data is highly non-normal. Our use of a t interval in (a) was not valid.

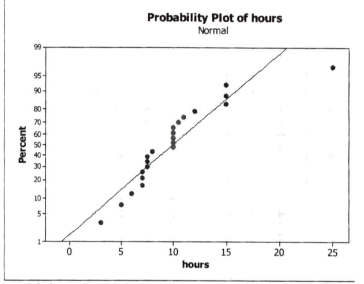

    c. In Minitab, put the data in C1 and execute the following macro 1000 times.

```
Let k3 = N(c1)
sample k3 c1 c3;
replace.
let k1=mean(c3)
stack k1 c5 c5
end
```

    d. Answers will vary; one bootstrap simulation gave $s_{boot} = 0.98$. The resulting 95% CI for μ is $9.955 \pm 2.080(0.98) = (7.917, 11.993)$. This is practically identical to the CI in (a).

    e. The bootstrap distribution still shows some positive skewness, which again makes the use of a t critical value somewhat suspect.

    f. Answers will vary; choosing the 25th bootstrap value from each end of the distribution in one simulation gave a percentile interval of (8.3409, 12.2045). Notice this is a right-ward shift from the other two intervals.

    g. Because of the non-normality of the bootstrap sampling distribution, we should not rely on t procedures; the bootstrap CI in (f) is preferred.

Chapter 8: Statistical Intervals Based on a Single Sample

51. For this data, $\bar{x} = 29.78$, $s = 13.07$, $n = 68$.

   a. $t_{.025,67} \approx 2$, so a 95% CI for μ is $29.78 \pm 2(13.07)/\sqrt{68} = (26.61, 32.94)$.

   b. Because of outliers, the weight gains do not seem normally distributed. However, with $n = 68$, the effects of the CLT might be enough to validate use of t procedures anyway.

   c. Use the macro provided for Exercise 49.

   d. Answers will vary; one bootstrap simulation gave $s_{boot} = 1.528$. The resulting 95% CI for μ is $29.78 \pm 2(1.528) = (26.72, 32.84)$. This is practically identical to the CI in (a).

   e. The bootstrap distribution look very normally distributed.

   f. Answers will vary; choosing the 25th bootstrap value from each end of the distribution in one simulation gave a percentile interval of $(26.88, 32.74)$. Notice this is virtually identical to the other two CIs.

   g. All three intervals are quite close, suggesting the use of a t CI in (a) was fine.

53. For this data, $\bar{x} = 38.65$, $s = 0.233$, $n = 8$.

   a. $t_{.025,7} = 2.365$, so a 95% CI for μ is $38.65 \pm 2.365(0.233)/\sqrt{8} = (38.46, 38.84)$.

   b. Although a normal probability plot is not perfectly straight, there is not enough deviation to reject normality.

   c. Use the macro provided for Exercise 49.

   d. Answers will vary; one bootstrap simulation gave $s_{boot} = 0.0762$. The resulting 95% CI for μ is $38.65 \pm 2.365(0.0762) = (38.47, 38.83)$. This is practically identical to the CI in (a).

   e. The bootstrap distribution is slightly right-skewed. The use of a t critical value in (d) might be suspect here.

   f. Answers will vary; choosing the 25th bootstrap value from each end of the distribution in one simulation gave a percentile interval of $(38.50, 38.81)$. Notice this is virtually identical to the other two CIs.

   g. All three intervals are quite close, suggesting the use of a t CI in (a) was fine.

   h. Normal body temperature is ~37 degrees Celsius, well below our CIs. This suggests the extreme conditions suffered by these Australian soldiers indeed influenced their body temperatures.

Chapter 8: Statistical Intervals Based on a Single Sample

55.   For this data, $\bar{x} = 187.28$, $s = 43.97$, $n = 25$.

    a. $t_{.025,24} = 2.064$, so a 95% CI for μ is $187.28 \pm 2.064(43.97)/\sqrt{25} = (169.13, 205.43)$.

    b.   A normal probability plot reveals highly non-normal data. With $n = 25$, this calls into question the t procedure in (a).

    c.   Use the macro provided for Exercise 49.

    d.   Answers will vary; one bootstrap simulation gave $s_{boot} = 8.35$. The resulting 95% CI for μ is $187.28 \pm 2.064(8.35) = (170.05,204.51)$.

    e.   The bootstrap distribution is slightly right-skewed. The use of a t critical value in (d) might be suspect here.

    f.   Answers will vary; choosing the 25th bootstrap value from each end of the distribution in one simulation gave a percentile interval of (173.20,205.64).

    g.   All three intervals contain the true $\mu = 178.29$. Of the three, we trust (f) the most, since (a) and (d) were both based on a dubious normality assumption.

57.

    a.   In Minitab, put the data in C1 and execute the following macro 1000 times. Results for the percentile interval will vary.

```
Let k3 = N(c1)
sample k3 c1 c3;
replace.
let k1=stdev(c3)
stack k1 c5 c5
end
```

    b.   Assuming normality, a 95% confidence interval for σ is (3.541, 6.578), but this interval is inappropriate because the normality assumption is clearly not satisfied.

59.

    a.   Assuming normality, $t_{.05,15} = 1.753$, so a 95% C.I. for μ is

$$.214 \pm 1.753 \frac{.036}{\sqrt{16}} = .214 \pm .016 = (.198,.230)$$

    b.   A 90% upper bound for σ, with $\chi^2_{.90,15} = 8.547$, is $\sqrt{\dfrac{15(.036)^2}{8.547}} = .0477$

    c.   A 95% prediction interval, with $t_{.025,15} = 2.131$, is

$$.214 \pm 2.131(.036)\sqrt{1 + \tfrac{1}{16}} = .214 \pm .0791 = (.1349,.2931).$$

Chapter 8: Statistical Intervals Based on a Single Sample

61. The specified condition is that the interval be length .2, so $n = \left[\dfrac{2(1.96)(.8)}{.2}\right]^2 = 245.86$,

so n = 246 should be used.

63.

   a. For this data, $\bar{x} = 0.16835$ and $s = 0.01938$. Assuming normality, $t_{.025,47} \approx 2.01$,

     so a 95% C.I. for μ is $0.16835 \pm 2.01\dfrac{0.01938}{\sqrt{48}} = (.163, .174)$. Yes, this interval

     is below the interval for 59(a).

   b. Nine of the 48 reaction times are below .15, so $\hat{p} = 9/48 = .1875$. Using the

     formula from this chapter, a 95% CI for the true p is $(.089, .326)$.

65. $\hat{p} = \dfrac{11}{55} = .2 \Rightarrow$ a 90% C.I. is

$$\dfrac{.2 + \dfrac{1.645^2}{2(55)} \pm 1.645\sqrt{\dfrac{(.2)(.8)}{55} + \dfrac{1.645^2}{4(55)^2}}}{1 + \dfrac{1.645^2}{55}} = \dfrac{.2246 \pm .0887}{1.0492} = (.1295, .2986)$$

67.

   a. A normal probability plot lends support to the assumption that pulmonary
     compliance is normally distributed. Note also that the lower and upper fourths
     are 192.3 and 228,1, so the fourth spread is 35.8, and the sample contains no
     outliers.

   b. $t_{.025,15} = 2.131$, so the C.I. is

$$209.75 \pm 2.131\dfrac{24.156}{\sqrt{16}} = 209.75 \pm 12.87 = (196.88, 222.62).$$

   c. k = 95, n = 16, and the tolerance critical value is 2.903, so the 95% tolerance
     interval is

$$209.75 \pm 2.903(24.156) = 209.75 \pm 70.125 = (139.625, 279.875).$$

Chapter 8: Statistical Intervals Based on a Single Sample

69.

a. The likelihood function is proportional to $\exp([2\sigma^2]^{-1}\sum(y_i - \beta x_i)^2)$. This is maximized exactly when $\sum(y_i - \beta x_i)^2$ is maximized, so differentiate this squared sum with respect to β and solve: $\sum 2(y_i - \beta x_i)(-x_i) = 0 \rightarrow -2[\sum y_i x_i - \beta \sum x_i^2] = 0$
$\rightarrow \beta = \sum y_i x_i / \sum x_i^2$. That is the MLE of beta is $\hat{\beta} = \sum x_i Y_i / \sum x_i^2$.

b. $E[\hat{\beta}] = E[\sum x_i Y_i / \sum x_i^2] = \sum x_i E[Y_i] / \sum x_i^2 = \sum x_i \beta x_i / \sum x_i^2 = \beta \sum x_i^2 / \sum x_i^2 = \beta$.

c. $V[\hat{\beta}] = V[\sum x_i Y_i / \sum x_i^2] = \sum x_i^2 V[Y_i] / (\sum x_i^2)^2 = \sum x_i^2 \sigma^2 / (\sum x_i^2)^2 = \sigma^2 / \sum x_i^2$. Thus, the standard deviation of $\hat{\beta}$ is $\sigma / \sqrt{\sum x_i^2}$.

d. The variance of $\hat{\beta}$ is inversely proportional to the sum of squares of the x values. Hence, to estimate $\hat{\beta}$ precisely – i.e., to reduce its variance – we should spread out the x values as far from zero as possible.

e. Solving the given expression for β gives the CI $\hat{\beta} \pm t_{.025, n-1} s / \sqrt{\sum x_i^2}$. By direct computation on these $n = 10$ values, we get a 95% CI of (29.93, 30.15).

71. The length of the interval is $\left(z_\gamma + z_{\alpha-\gamma}\right)\dfrac{s}{\sqrt{n}}$, which is minimized when $z_\gamma + z_{\alpha-\gamma}$ is minimized, i.e. when $\Phi^{-1}(1-\gamma) + \Phi^{-1}(1-\alpha+\gamma)$ is minimized. Taking $\dfrac{d}{d\gamma}$ and equating to 0 yields $\dfrac{1}{\Phi(1-\gamma)} = \dfrac{1}{\Phi(1-\alpha+\gamma)}$ where $\Phi(\bullet)$ is the standard normal p.d.f., whence $\gamma = \dfrac{\alpha}{2}$.

73.

a. Since $2\lambda\sum X_i$ has a chi-squared distribution with 2n d.f. and the area under this chi-squared curve to the right of $\chi^2_{.95, 2n}$ is .95, $P\left(\chi^2_{.95, 2n} < 2\lambda\sum X_i\right) = .95$. This implies that $\dfrac{\chi^2_{.95, 2n}}{2\sum X_i}$ is a lower confidence bound for λ with confidence coefficient 95%. Table A.7 gives the chi-squared critical value for 20 d.f. as

Chapter 8: Statistical Intervals Based on a Single Sample

10.851, so the bound is $\dfrac{10.851}{2(550.87)} = .0098$. We can be 95% confident that λ exceeds .0098.

b. Arguing as in a, $P\left(2\lambda\Sigma X_i < \chi^2_{.05,2n}\right) = .95$. The following inequalities are equivalent to the one in parentheses:

$$\lambda < \frac{\chi^2_{.05,2n}}{2\Sigma X_i} \Rightarrow -\lambda t < \frac{-t\chi^2_{.05,2n}}{2\Sigma X_i} \Rightarrow e^{-\lambda t} < \exp\left[\frac{-t\chi^2_{.05,2n}}{2\Sigma X_i}\right].$$

Replacing the ΣX_i by Σx_i in the expression on the right hand side of the last inequality gives a 95% lower confidence bound for $e^{-\lambda t}$. Substituting $t = 100$, $\chi^2_{.05,20} = 31.410$ and $\Sigma x_i = 550.87$ gives .058 as the lower bound for the probability that time until breakdown exceeds 100 minutes.

75.

a. Let the $100p$th percentile of a standard normal population be z, so the corresponding percentile for a general normal population is $\theta = \mu + \sigma z$, or $z = (\theta - \mu)/\sigma$. Substitute this into the given expression for t, and cancel: $t =$

$$\frac{\sigma}{s}\left[\frac{\bar{x}-\mu}{\sigma/\sqrt{n}} - \left(\frac{\theta-\mu}{\sigma}\right)\sqrt{n}\right] = \frac{\bar{x}-\theta}{s/\sqrt{n}},$$ so this quantity has a non-central t distribution. Set this equal to the appropriate critical values (those provided), and a 95% CI for θ is $(\bar{x} - (s/\sqrt{n})t_{.025,n-1,\delta}, \bar{x} - (s/\sqrt{n})t_{.975,n-1,\delta})$.

b. The 5th percentile of the Z distribution is $z = -1.645$, so $\delta = -(-1.645)\sqrt{16} = 6.58$. Hence, we can use the critical values provided. Using the data from Example 8.11 ultimately gives a 95% CI of $(3.01, 4.46)$.

77.

a. Using independence, the probability is still just $(1/2)(1/2)\ldots(1/2) = (1/2)^n$.

b. There are n possible X_i's that could fall below the median, so the probability exactly one does is n times the answer from (a), or $n(1/2)^n$.

    c.   The event $Y_2 > \tilde{\mu}$ occurs if either all n observations lie above the median or exactly one of the observations lies below the median. Combining the previous exercise and (b), $P(Y_2 > \tilde{\mu}) = (1/2)^n + n(1/2)^n = (n+1)(1/2)^n$. Using the same rationale as the previous exercise, the CI (y_2, y_{n-1}) for $\tilde{\mu}$ has confidence level $1 - 2[(n+1)(1/2)^n] = 1 - (n+1)(1/2)^{n-1}$. For $n = 10$, this confidence level is 97.85%, and the CI from the data is $(y_2, y_{n-1}) = (29.9, 39.3)$.

79.

    a.   If A_1 and A_2 are independent, then $P(A_1 \cap A_2) = P(A_1)P(A_2) = (.95)^2 = 90.25\%$.

    b.   For any events A and B, $P(A \text{ or } B) = P(A) + P(B) - P(A \cap B) \le P(A) + P(B)$. Apply that here: $P(A_1' \text{ or } A_2') \le P(A_1') + P(A_2') = (1-.95) + (1-.95) = .10$, so that $P(A_1 \cap A_2) = 1 - P(A_1' \text{ or } A_2') \ge 1 - .10 = .90$.

    c.   Replace .05 with α above, and you find $P(A_1 \cap A_2) \ge 100(1 - 2\alpha)\%$. In general, the simultaneous confidence level for k separate CIs is at least $100(1 - k\alpha)\%$.

Chapter 9: Tests of Hypotheses Based on a Single Sample

1.

    a. Yes. It is an assertion about the value of a parameter.

    b. No. The sample median $\widetilde{X}$ is not a parameter.

    c. No. The sample standard deviation s is not a parameter.

    d. Yes. The assertion is that the standard deviation of population #2 exceeds that of population #1

    e. No. $\overline{X}$ and $\overline{Y}$ are statistics rather than parameters, so cannot appear in a hypothesis.

    f. Yes. H is an assertion about the value of a parameter.

3.   In this formulation, H_o states the welds do not conform to specification. This assertion will not be rejected unless there is strong evidence to the contrary. Thus the burden of proof is on those who wish to assert that the specification is satisfied. Using H_a: $\mu < 100$ results in the welds being believed in conformance unless proved otherwise, so the burden of proof is on the non-conformance claim.

5.   Let σ denote the population standard deviation. The appropriate hypotheses are $H_o : \sigma = .05$ vs $H_a : \sigma < .05$. With this formulation, the burden of proof is on the data to show that the requirement has been met (the sheaths will not be used unless H_o can be rejected in favor of H_a. Type I error: Conclude that the standard deviation is < .05 mm when it is really equal to .05 mm. Type II error: Conclude that the standard deviation is .05 mm when it is really < .05.

7.   A type I error here involves saying that the plant is not in compliance when in fact it is. A type II error occurs when we conclude that the plant is in compliance when in fact it isn't. Reasonable people may disagree as to which of the two errors is more serious. If in your judgement it is the type II error, then the reformulation $H_o : \mu = 150$ vs $H_a : \mu < 150$ makes the type I error more serious.

9.

    a. R_1 is most appropriate, because x either too large or too small contradicts p = .5 and supports p $\neq$ .5.

    b. A type I error consists of judging one of the two companies favored over the other when in fact there is a 50-50 split in the population. A type II error involves judging the split to be 50-50 when it is not.

c. X has a binomial distribution with n = 25 and p = 0.5. α = P(type I error) = $P(X \leq 7 \text{ or } X \geq 18$ when $X \sim$ Bin(25, .5)) = B(7; 25,.5) + 1 − B(17; 25,.5) = .044

d. $\beta(.4) = P(8 \leq X \leq 17$ when p = .4) = B(17; 25,.5) − B(7, 25,.4) = 0.845, and $\beta(.6) = 0.845$ also. $\beta(.3) = B(17;25,.3) − B(7;25,.3) = .488 = \beta(.7)$

e. x = 6 is in the rejection region R_1 , so H_o is rejected in favor of H_a.

11.

a. $H_o : \mu = 10$ vs $H_a : \mu \neq 10$

b. $\alpha = P($ rejecting H_o when H_o is true$) = P(\bar{x} \geq 10.1032$ or $\leq 9.8968 \text{ when } \mu = 10)$. Since $\bar{x}$ is normally distributed with standard deviation
$$\frac{\sigma}{\sqrt{n}} = \frac{.2}{5} = .04, \ \alpha = P(z \geq 2.58 \text{ or } \leq -2.58) = .005 + .005 = .01$$

c. When $\mu = 10.1$, $E(\bar{x}) = 10.1$, so $\beta(10.1) = P(9.8968 < \bar{x} < 10.1032$ when $\mu = 10.1) = P(-5.08 < z < .08) = .5319$. Similarly, $\beta(9.8) = P(2.42 < z < 7.58) = .0078$

d. $c = \pm 2.58$

e. Now $\frac{\sigma}{\sqrt{n}} = \frac{.2}{3.162} = .0632$. Thus 10.1032 is replaced by c, where $\frac{c - 10}{.0632} = 1.96$ and so c = 10.124. Similarly, 9.8968 is replaced by 9.876.

f. $\bar{x} = 10.020$. Since $\bar{x}$ is neither ≥ 10.124 nor ≤ 9.876, it is not in the rejection region. H_o is not rejected; it is still plausible that $\mu = 10$.

g. $\bar{x} \geq 10.1032$ or ≤ 9.8968 iff $z \geq 2.58$ or ≤ -2.58.

Chapter 9: Tests of Hypotheses Based on a Single Sample

13.

a. $P(\bar{x} \geq \mu_o + 2.33\frac{\sigma}{\sqrt{n}}\ when\mu = \mu_o) = P\left(Z \geq \dfrac{\left(\mu_o + 2.33\dfrac{\sigma}{\sqrt{n}} - \mu_o\right)}{\dfrac{\sigma}{\sqrt{n}}}\right)$

$= P(Z \geq 2.33) = .01$, where Z is a standard normal r.v.

b. P(rejecting H_o when $\mu = 99$) = $P(\bar{x} \geq 102.33$ when $\mu = 99)$

$= P\left(z \geq \dfrac{102 - 99}{1}\right) = P(z \geq 3.33) = .0004$. Similarly,

$\alpha(98) = P(\bar{x} \geq 102.33$ when $\mu = 98) = P(z \geq 4.33) = 0$. In general, we have P(type I error) < .01 when this probability is calculated for a value of μ less than 100. The boundary value $\mu = 100$ yields the largest α.

15.

a. $\alpha = P(z \geq 1.88$ when z has a standard normal distribution)
$= 1 - \Phi(1.88) = .0301$

b. $\alpha = P(z \leq -2.75$ when z ~ N(0, 1) $= \Phi(-2.75) = .003$

c. $\alpha = \Phi(-2.88) + (1 - \Phi(2.88)) = .004$

17.

a. $z = \dfrac{30,960 - 30,000}{1500 / \sqrt{16}} = 2.56 > 2.33$ so reject H_o.

b. $\beta(30,500): \Phi\left(2.33 + \dfrac{30,000 - 30,500}{1500 / \sqrt{16}}\right) = \Phi(1.00) = .8413$

c. $\beta(30,500) = .05 : n = \left[\dfrac{1500(2.33 + 1.645)}{30,000 - 30,500}\right]^2 = 142.2$, so use n = 143

d. $\alpha = 1 - \Phi(2.56) = .0052$

Chapter 9: Tests of Hypotheses Based on a Single Sample

19.

   a. Reject H_o if either $z \geq 2.58$ or $z \leq -2.58$; $\dfrac{\sigma}{\sqrt{n}} = 0.3$, so

$$z = \frac{94.32 - 95}{0.3} = -2.27. \text{ Since } -2.27 \text{ is not} < -2.58, \text{ don't reject } H_o.$$

   b. $\beta(94) = \Phi\left(2.58 + \dfrac{1}{0.3}\right) - \Phi\left(-2.58 + \dfrac{1}{0.3}\right) = \Phi(5.91) - \Phi(.75) = .2266$

   c. $n = \left[\dfrac{1.20(2.58 + 1.28)}{95 - 94}\right]^2 = 21.46$, so use n = 22.

21. With H_o: $\mu = .5$, and H_a: $\mu \neq .5$ we reject H_o if $t > t_{\alpha/2, n-1}$ or $t < -t_{\alpha/2, n-1}$

   a. $1.6 < t_{.025, 12} = 2.179$, so don't reject H_o

   b. $-1.6 > -t_{.025, 12} = -2.179$, so don't reject H_o

   c. $-2.6 > -t_{.005, 24} = -2.797$, so don't reject H_o

   d. $-3.9 <$ the negative of all t values in the df = 24 row, so we reject H_o in favor of H_a.

23. H_o: $\mu = 360$ vs. H_a: $\mu > 360$; $t = \dfrac{\bar{x} - 360}{s/\sqrt{n}}$; reject H_o if $t > t_{.05, 25} = 1.708$;

$$t = \frac{370.69 - 360}{24.36/\sqrt{26}} = 2.24 > 1.708. \text{ Thus } H_o \text{ should be rejected. There appears to be a}$$

contradiction of the prior belief.

25. To test H_o: $\mu = 100$ vs. H_a: $\mu \neq 100$, with a normal population and $\sigma = 15$ <u>known</u>, we

use a z-test and reject H_o if $|z| > 1.96$. Here, $z = \dfrac{\bar{x} - 100}{15/\sqrt{n}} = \dfrac{116 - 100}{15/\sqrt{10}} = 3.37$, so we

certainly reject H_0. We conclude that the mean IQ for all first graders at this school is significantly different than the national average of 100.

Chapter 9: Tests of Hypotheses Based on a Single Sample

27.

a. We will test H_o: $\mu \geq 113$ vs. H_a: $\mu < 113$ and reject H_o if $t < -t_{.05,5} = -2.015$. Here,

$t = \dfrac{\bar{x} - 113}{s / \sqrt{n}} = \dfrac{112.97 - 113}{4.29 / \sqrt{6}} = -.02$. We fail to reject H_0 here; there is no

significant evidence that the population mean isn't at least 113 grams.

b. Under these assumptions, $\bar{X}$ is normally distributed with mean $\mu = 110$ and

standard error $4 / \sqrt{6} = 1.633$. In this one-sided z test, we would reject H_o if $z < -$

1.645, where $z = \dfrac{\bar{x} - 113}{4 / \sqrt{6}}$. The probability of rejection is, thus,

$$P\left(\dfrac{\bar{X} - 113}{4 / \sqrt{6}} < -1.645 \right) = P\left(\dfrac{\bar{X} - 110}{4 / \sqrt{6}} < -1.645 + \dfrac{113 - 110}{4 / \sqrt{6}} \right) = P(Z < 0.19) = .58.$$

c. Replace 6 with n in the last part of (b): we want

$$P\left(Z < -1.645 + \dfrac{113 - 110}{4 / \sqrt{n}} \right) \geq .95. \text{ This requires}$$

$-1.645 + \dfrac{113 - 110}{4 / \sqrt{n}} > \Phi^{-1}(.95) = 1.645$, which solving for n gives $n > 19.24$. So,

at least 20 observations are required total, a.k.a. an additional 14 observations.

29.

a. For n = 8, n − 1 = 7, and $t_{.05,7} = 1.895$, so H_o is rejected at level .05 if

$t \geq 1.895$. Since $\dfrac{s}{\sqrt{n}} = \dfrac{1.25}{\sqrt{8}} = .442$, $t = \dfrac{3.72 - 3.50}{.442} = .498$; this does

not exceed 1.895, so H_o is not rejected.

b. $d = \dfrac{|\mu_o - \mu|}{\sigma} = \dfrac{|3.50 - 4.00|}{1.25} = .40$, and df = 7, so from table A.17, $\beta(.40) \approx .73$

31. The hypotheses of interest are H_o: $\mu = 7$ vs H_a: $\mu < 7$, so a lower-tailed test is

appropriate; H_o should be rejected if $t \leq -t_{.1,8} = -1.397$. $t = \dfrac{6.32 - 7}{1.65 / \sqrt{9}} = -1.24$.

Because -1.24 is not ≤ -1.397, H_o (prior belief) is not rejected (contradicted) at
level .01.

Chapter 9: Tests of Hypotheses Based on a Single Sample

33.

$$\beta(\mu_o - \Delta) = \Phi\!\left(z_{\alpha/2} + \Delta\sqrt{n}\,/\,\sigma\right) - \Phi\!\left(-z_{\alpha/2} + \Delta\sqrt{n}\,/\,\sigma\right)$$
$$= 1 - \Phi\!\left(-z_{\alpha/2} - \Delta\sqrt{n}\,/\,\sigma\right) - [1 - \Phi\!\left(z_{\alpha/2} - \Delta\sqrt{n}\,/\,\sigma\right)] =$$
$$\Phi\!\left(z_{\alpha/2} - \Delta\sqrt{n}\,/\,\sigma\right) - \Phi\!\left(-z_{\alpha/2} - \Delta\sqrt{n}\,/\,\sigma\right) = \beta(\mu_o + \Delta)$$

35.

a.   A boxplot of the data is quite symmetric with no outliers, suggesting a normally distributed population is plausible.

b.   We will test H_o: $\mu = 1.5$ vs. H_a: $\mu \neq 1.5$ and reject H_o if $|t| > t_{.025,7} = 2.365$. Here,
$$t = \frac{1.816 - 1.5}{0.2105/\sqrt{8}} = 4.25 \, .$$ So, we definitely reject H_o and conclude the true average amount poured differs significantly from 1.5 oz.

c.   Yes: with $n = 8$ and σ unknown, the use of the t test relies on a normally distributed population. A computer-generated probability plot confirms what we noted in (a): a normal population is very plausible for this data.

d.   The tolerance critical value for 95% coverage and 95% confidence with $n = 8$ is 3.732 from Table A.6. So, for a confidence level of 95%, an interval for capturing at least 95% of the amounts is $1.816 \pm 3.732(0.2105) = (1.03, 2.60)$.

37.

a.

1	p = true proportion of all nickel plates that blister under the given circumstances.
2	H_o: $p = .10$
3	H_a: $p > .10$
4	$z = \dfrac{\hat{p} - p_o}{\sqrt{p_o(1 - p_o)/n}} = \dfrac{\hat{p} - .10}{\sqrt{.10(.90)/n}}$
5	Reject H_o if $z \geq 1.645$
6	$z = \dfrac{14/100 - .10}{\sqrt{.10(.90)/100}} = 1.33$
7	Fail to Reject H_o. The data does not give compelling evidence for concluding that more than 10% of all plates blister under the circumstances.

The possible error we could have made is a Type II error: Failing to reject the null hypothesis when it is actually true.

b. $\beta(.15) = \Phi\left[\dfrac{.10 - .15 + 1.645\sqrt{.10(.90)/100}}{\sqrt{.15(.85)/100}}\right] = \Phi(-.02) = .4920$. When n = 200,

$\beta(.15) = \Phi\left[\dfrac{.10 - .15 + 1.645\sqrt{.10(.90)/200}}{\sqrt{.15(.85)/200}}\right] = \Phi(-.60) = .2743$

c. $n = \left[\dfrac{1.645\sqrt{.10(.90)} + 1.28\sqrt{.15(.85)}}{.15 - .10}\right]^2 = 19.01^2 = 361.4$, so use n = 362

39.

a. We wish to test H_0: p = .02 vs H_a: p < .02; only if H_0 can be rejected will the inventory be postponed. The lower-tailed test rejects H_0 if z ≤ -1.645. With $\hat{p} = \dfrac{15}{1000} = .015$, z = -1.01, which is not ≤ -1.645. Thus, H_0 cannot be rejected, so the inventory should be carried out.

b. $\beta(.01) = 1 - \Phi\left[\dfrac{.02 - .01 - 1.645\sqrt{.02(.98)/1000}}{\sqrt{.01(.99)/1000}}\right] = 1 - \Phi(0.86) = .1949$

c. $\beta(.05) = 1 - \Phi\left[\dfrac{.02 - .05 - 1.645\sqrt{.02(.98)/1000}}{\sqrt{.05(.95)/1000}}\right] = 1 - \Phi(-5.41) \approx 1$, so the chance the inventory will be *postoned* is P(reject H_0 when p = .05) = 1 - $\beta(.05)$ = 0. It is highly unlikely that H_0 will be rejected, and the inventory will almost surely be carried out.

41.

a. p = true proportion of current customers who qualify. H_0: p = .05 vs H_a: p ≠ .05, $z = \dfrac{\hat{p} - .05}{\sqrt{.05(.95)/n}}$, reject H_0 if z ≥ 2.58 or z ≤ -2.58. $\hat{p} = .08$, so $z = \dfrac{.03}{.00975} = 3.07 \geq 2.58$, so H_0 is rejected. The company's premise is not correct.

b. $\beta(.10) = \Phi\left[\dfrac{.05 - .10 + 2.58\sqrt{.05(.95)/500}}{\sqrt{.10(.90)/500}}\right] - \Phi\left[\dfrac{.05 - .10 - 2.58\sqrt{.05(.95)/500}}{\sqrt{.10(.90)/500}}\right]$

$\approx \Phi(-1.85) - 0 = .0332$

Chapter 9: Tests of Hypotheses Based on a Single Sample

43.   The hypotheses are H_o: $p = .10$ vs. H_a: $p > .10$, so R has the form $\{c, \ldots, n\}$. The values $n = 10$, $c = 3$ (i.e. $R = \{3, 4, \ldots, 10\}$) yield $\alpha = 1 - B(2; 10, .1) = .07$ while no larger R has $\alpha \le .10$; however $\beta(.3) = B(2; 10, .3) = .383$. For $n = 20$, $c = 5$ yields $\alpha = 1 - B(4; 20, .1) = .043$, but again $\beta(.3) = B(4; 20, .3) = .238$. For $n = 25$, $c = 5$ yields $\alpha = 1 - B(4; 25, .1) = .098$ while $\beta(.7) = B(4; 25, .3) = .090 \le .10$, so $n = 25$ should be used. The rejection region is $R = \{5, \ldots, 25\}$, $\alpha = .098$, and $\beta(.7) = .090$.

45.   Using $\alpha = .05$, H_o should be rejected whenever P-value $< .05$.
   a.   P-value $= .001 < .05$, so reject H_o
   b. $.021 < .05$, so reject H_o.
   c. $.078$ is not $< .05$, so don't reject H_o.
   d. $.047 < .05$, so reject H_o (a close call).
   e. $.148 > .05$, so H_o can't be rejected at level .05.

47.   In each case the p-value $= P(Z > z) = 1 - \Phi(z)$.
   a.   .0778
   b.   .1841
   c.   .0250
   d.   .0066
   e.   .5438

49.   Use Table A.8.
   a. $P(t > 2.0)$ at 8df $= .040$
   b. $P(t < -2.4)$ at 11df $= .018$
   c. $2P(t < -1.6)$ at 15df $= 2(.065) = .130$
   d.   by symmetry, $P(t > -.4) = 1 - P(t > .4)$ at 19df $= 1 - .347 = .653$
   e. $P(t > 5.0)$ at 5df $< .005$
   f. $2P(t < -4.8)$ at 40df $< 2(.000) = .000$ to three decimal places

51.   The p-value is greater than the level of significance $\alpha = .01$, therefore fail to reject H_o that $\mu = 5.63$. The data does not indicate a statistically significant difference in average serum receptor concentration between pregnant women and all other women.

53.
   a.   For testing H_o: $p = .2$ vs H_a: $p > .2$, an upper-tailed test is appropriate. The computed Z is $z = .97$, so p-value $= 1 - \Phi(.97) = .166$. Because the p-value is rather large, H_o would not be rejected at any reasonable α (it can't be rejected for any $\alpha < .166$), so no modification appears necessary.

   b.   With $p = .5$, $1 - \beta(.5) = 1 - \Phi[(-.3 + 2.33(.0516))/.0645] = 1 - \Phi(-2.79) = .9974$

Chapter 9: Tests of Hypotheses Based on a Single Sample

55. μ = the true average percentage of organic matter in this type of soil, and the hypotheses are H_o: $\mu = 3$ vs H_a: $\mu \neq 3$. With n = 30, and assuming normality, we use the t test: $t = \dfrac{\bar{x} - 3}{s/\sqrt{n}} = \dfrac{2.481 - 3}{.295} = \dfrac{-.519}{.295} = -1.759$. The p-value = 2[P(t > 1.759)] = 2(.041) = .082. At significance level .10, since $.082 \leq .10$, we would reject H_0 and conclude that the true average percentage of organic matter in this type of soil is something other than 3. At significance level .05, we would not have rejected H_0.

57.

    a.   The appropriate hypotheses are H_o: $\mu = 10$ vs H_a: $\mu < 10$

    b.   P-value = P(t > 2.3) = .017, which is $\leq .05$, so we would reject H_o. The data indicates that the pens do not meet the design specifications.

    c.   P-value = P(t > 1.8) = .045, which is not $\leq .01$, so we would not reject H_o. There is not enough evidence to say that the pens don't satisfy the design specifications.

    d.   P-value = P(t > 3.6) $\approx .001$, which gives strong evidence to support the alternative hypothesis.

59. With H_o: $\mu = .60$ vs H_a: $\mu \neq .60$,and a two-tailed p-value of .0711, we fail to reject H_o at levels .01 and .05 (thus concluding that the amount of impurities need not be adjusted) , but we would reject H_o at level .10 (and conclude that the amount of impurities does need adjusting).

61.

    a.   Here $\beta = \Phi\left(\dfrac{-.01 + .9320/\sqrt{n}}{.4073/\sqrt{n}}\right) = \Phi\left(\dfrac{\left(-.01\sqrt{n} + .9320\right)}{.4073}\right) = .9793, .8554, .4325,$ .0944, and 0 for n = 100, 2500, 10,000, 40,000, and 90,000, respectively.

    b.   Here $z = .025\sqrt{n}$ which equals .25, 1.25, 2.5, and 5 for the four n's, whence p-value = .4213, .1056, .0062, .0000, respectively.

    c.   No; the reasoning is the same as in 55 (c).

Chapter 9: Tests of Hypotheses Based on a Single Sample

63.

a. The likelihood function here is $f(\mathbf{x};\sigma^2) = (2\pi\sigma^2)^{-n/2} \exp(-\Sigma x_i^2 / 2\sigma^2)$, so the most powerful test rejects H_0 when $\dfrac{(2\pi(3)^2)^{-n/2} \exp(-\Sigma x_i^2 / 2(3)^2)}{(2\pi(2)^2)^{-n/2} \exp(-\Sigma x_i^2 / 2(2)^2)} \geq k$, a.k.a.

$\exp(\Sigma x_i^2 / 2(2)^2 - \Sigma x_i^2 / 2(3)^2) \geq k'$. Taking logarithms to solve for $\sum x_i^2$ gives a solution of the form $\sum x_i^2 \geq c$.

b. Under normality with μ known, $\sum x_i^2 \sim \sigma^2 \chi_{10}^2$ when $n = 10$. Hence, we want $.05$ $= \alpha = P(\sum x_i^2 \geq c$ when $\sigma^2 = 2) = P(2 \chi_{10}^2 \geq c) = P(\chi_{10}^2 \geq c/2) \rightarrow c = 2 \chi_{.05,10}^2 = 2(18.307) = 36.614$.

c. Yes – the Neyman-Pearson method in (a) would yield the same test form (i.e., reject the null if $\sum x_i^2 \geq c$) for any alternative value $\sigma_a^2 > 2$, not just 3.

65.

a. The likelihood function here is $f(\mathbf{x};\theta) = \lambda^n \exp(-\lambda\Sigma x_i)$, so the most powerful test rejects H_0 when $\dfrac{(.5)^n \exp(-(.5)\Sigma x_i)}{1^n \exp(-1\Sigma x_i)} \geq k$, a.k.a. $\exp(.5\Sigma x_i) \geq k'$. Taking logarithms gives a solution of the form $\sum x_i \geq c$.

b. Yes – the Neyman-Pearson method in (a) would yield the same test form (i.e., reject the null if $\sum x_i \geq c$ for some c) for any alternative value $\lambda < 1$, not just 0.5.

67. Rewrite the likelihood as $C\theta^{2x_1+x_2}(1-\theta)^{x_2+2x_3} = C\theta^{2x_1+x_2}(1-\theta)^{2n-[2x_1+x_2]} = C\theta^y(1-\theta)^{2n-y}$, where $y = 2x_1 + x_2$. Then the most powerful test rejects H_0 when $\dfrac{C(.8)^y(1-.8)^{2n-y}}{C(.5)^y(1-.5)^{2n-y}} \geq k$, a.k.a. $\left(\dfrac{.8}{1-.8}\right)^y \geq k'$. Solving for y gives a solution of the form $y \geq c$. Yes, the test is UMP for the alternative $H_a : \theta > .5$ because the tests for H_0 : $\theta = .5$ vs. $H_a : \theta = p_0$ all have the same form for $p_0 > .5$.

Chapter 9: Tests of Hypotheses Based on a Single Sample

69.

a. The usual test rejects H_0 at $\alpha = .05$ if $|z| > z_{.025} = 1.96$, where

$$z = \frac{\bar{x} - \mu_0}{\sigma / \sqrt{n}} = \frac{\bar{x} - 0}{4 / \sqrt{16}} = \bar{x}.$$

b. $\pi(0) = P(\bar{X} \geq 2.17 \text{ or } \bar{X} \leq -1.81 \text{ when } \mu = 0) = P(Z \geq 2.17 \text{ or } Z \leq -1.81) = 1 - \Phi(2.17) + \Phi(-1.81) = .0502.$

c. $\pi(.1) = P(\bar{X} \geq 2.17 \text{ or } \bar{X} \leq -1.81 \text{ when } \mu = .1) = P(Z \geq 2.07 \text{ or } Z \leq -1.91) = 1 - \Phi(2.07) + \Phi(-1.91) = .04345.$ Similarly, $\pi(-.1) = 1 - \Phi(2.27) + \Phi(-1.71) = .05826.$ This test is *not* unbiased, since $\pi(.1) < \pi(0)$. By definition, a test is unbiased only if its power is greater on the alternative than on the null.

d. $\pi(.1) = P(\bar{X} \geq 1.96 \text{ or } \bar{X} \leq -1.96 \text{ when } \mu = .1) = 1 - \Phi(1.86) + \Phi(-2.06) = .05114.$ Similarly, $\pi(-.1) = 1 - \Phi(2.06) + \Phi(-1.86) = .05114.$ This test is not UMP, since it has less power for the alternative $\mu = -.1$ than the test in (b) and (c): $.05114 < .05286.$

71.

a. From the algebra already presented in the text, $\lambda =$

$$\left(\frac{1}{1 + n(\bar{x} - \mu_0)^2 / \sum(x_i - \bar{x})^2} \right)^{n/2} =$$

$$\left(1 + \frac{n(\bar{x} - \mu_0)^2}{\sum(x_i - \bar{x})^2} \right)^{-n/2} = \left(1 + \frac{n(\bar{x} - \mu_0)^2}{(n-1)s^2} \right)^{-n/2} = \left(1 + \frac{[(\bar{x} - \mu_0)/(s/\sqrt{n})]^2}{n-1} \right)^{-n/2}$$

$$= \left(1 + \frac{t^2}{n-1} \right)^{-n/2}.$$ The approximate chi-square statistic is then $-2\ln(\lambda) = -2(-$

$$n/2)\ln\left(1 + \frac{t^2}{n-1} \right) = n \ln\left(1 + \frac{t^2}{n-1} \right).$$

b. In exercise 55, we found $t = -1.759$. Substitute t and $n = 30$ into the above expression to get $-2\ln(\lambda) = 3.041$. $-2\ln(\lambda)$ has a chi-square distribution with $v = 2 - 1 = 1$, so the *P*-value is .081, compared to .089 for Exercise 55.

73. Here we assume that thickness is normally distributed, so that for any n a t test is appropriate, and use Table A.17 to determine n. We wish $\beta(3) = .05$ when

$$d = \frac{|3.2 - 3|}{.3} = .667.$$ By inspection, df $= 29$ (n $= 30$) satisfies this requirement, so n $=$ 50 is unnecessarily large.

155

Chapter 9: Tests of Hypotheses Based on a Single Sample

75.

    a. H_o: $\mu = 2150$ vs H_a: $\mu > 2150$

    b. $t = \dfrac{\bar{x} - 2150}{s / \sqrt{n}}$

    c. $t = \dfrac{2160 - 2150}{30 / \sqrt{16}} = \dfrac{10}{7.5} = 1.33$

    d. At 15df, P-value $= P(t > 1.33) = .107$ (approximately)

    e. From **d**, p-value $> .05$, so H_o cannot be rejected at this significance level. The mean tensile strength for springs made using roller straightening is not significantly greater than 2150 N/mm^2.

77. $n = 8, \bar{x} = 30.7875, s = 6.5300$

    1   Parameter of interest: μ = true average heat-flux of plots covered with coal dust

    2   H_o: $\mu = 29.0$

    3   H_a: $\mu > 29.0$

    4       $t = \dfrac{\bar{x} - 29.0}{s / \sqrt{n}}$

    5   RR: $t \geq t_{\alpha, n-1}$ or $t \geq 1.895$

    6       $t = \dfrac{30.7875 - 29.0}{6.53 / \sqrt{8}} = .7742$

    7   Fail to reject H_o. The data does not indicate the mean heat-flux for pots covered with coal dust is greater than for plots covered with grass.

79.   At the .05 significance level, reject H_o because $.043 < .05$. At the level .01, fail to reject H_o because $.043 > .01$. Thus the data contradicts the design specification that sprinkler activation is less than 25 seconds at the level .05, but not at the .01 level.

81.   A normality plot reveals that these observations could have come from a normally distributed population, therefore a t-test is appropriate. The relevant hypotheses are H_o: $\mu = 9.75$ vs H_a: $\mu > 9.75$. Summary statistics are n = 20, $\bar{x} = 9.8525$, and s = .0965, which leads to a test statistic $t = \dfrac{9.8525 - 9.75}{.0965 / \sqrt{20}} = 4.75$, from which the p-value = .0001. (From MINITAB output). With such a small p-value, the data strongly supports the alternative hypothesis. The condition is not met.

Chapter 9: Tests of Hypotheses Based on a Single Sample

83. A t test is appropriate; H_o: $\mu = 1.75$ is rejected in favor of H_a: $\mu \neq 1.75$ if the P-value $< .05$. The computed t is $t = \dfrac{1.89 - 1.75}{.42/\sqrt{26}} = 1.70$. Since the P-value is $2P(t > 1.7)$ $= 2(.051) = .102 > .05$, do not reject H_o; the data does not contradict prior research. We assume that the population from which the sample was taken was approximately normally distributed.

85. Let p = the true proportion of mechanics who could identify the problem. Then the appropriate hypotheses are H_o: p = .75 vs H_a: p < .75, so a lower-tailed test should be used. With p_o= .75 and $\hat{p} = \dfrac{42}{72} = .583$, z = -3.28 and $P = \Phi(-3.28) = .0005$.

Because this p-value is so small, the data argues strongly against H_o, so we reject it in favor of H_a.

87. H_o: $\mu = 15$ vs H_a: $\mu > 15$. Because the sample size is less than 40, and we can assume the distribution is approximately normal, the appropriate statistic is $t = \dfrac{\overline{x} - 15}{s/\sqrt{n}} = \dfrac{17.5 - 15}{2.2/\sqrt{32}} = \dfrac{2.5}{.390} = 6.4$. Thus the p-value is "off the chart" in the 20 df column of Table A.8, and so is approximately $0 < .05$, so H_o is rejected in favor of the conclusion that the true average time exceeds 15 minutes.

89. The 20 df row of Table A.7 shows that $\chi^2_{.99,20} = 8.26 < 8.58$ (H_o not rejected at level .01) and $8.58 < 9.591 = \chi^2_{.975,20}$ (H_o rejected at level .025). Thus $.01 < $ p-value $< .025$ and H_o cannot be rejected at level .01 (the p-value is the smallest alpha at which rejection can take place, and this exceeds .01).

91.

a. When H_o is true, $2\lambda_o \Sigma X_i = 2\sum \dfrac{X_i}{\mu_o}$ has a chi-squared distribution with df = 2n.

If the alternative is H_a: $\mu > \mu_o$, large test statistic values (large Σx_i, since $\overline{x}$ is large) suggest that H_o be rejected in favor of H_a, so rejecting when $2\sum \dfrac{X_i}{\mu_o} \geq \chi^2_{\alpha,2n}$ gives a test with significance level α. If the alternative is H_a: $\mu < \mu_o$, rejecting when $2\sum \dfrac{X_i}{\mu_o} \leq \chi^2_{1-\alpha,2n}$ gives a level α test. The rejection region for H_a: $\mu \neq \mu_o$ is either $2\sum \dfrac{X_i}{\mu_o} \geq \chi^2_{\alpha/2,2n}$ or $\leq \chi^2_{1-\alpha/2,2n}$.

b. H_o: $\mu = 75$ vs H_a: $\mu < 75$. The test statistic value is $\dfrac{2(737)}{75} = 19.65$. At

level .01, H_o is rejected if $2\sum \dfrac{X_i}{\mu_o} \le \chi^2_{.99,20} = 8.260$. Clearly 19.65 is not in the

rejection region, so H_o should not be rejected. The sample data does not suggest
that true average lifetime is less than the previously claimed value.

93.

a. $\alpha = P(X \le 5$ when $p = .9) = B(5; 10, .9) = .002$, so the region $(0, 1, ..., 5)$ does
specify a level .01 test.

b. The first value to be placed in the upper-tailed part of a two tailed region would
be 10, but $P(X = 10$ when $p = .9) = .349$, so whenever 10 is in the rejection
region, $\alpha \ge .349$.

c. $\beta(p') = P(X$ is <u>not</u> in R when $p = p') = P(X > 5$ when $p = p') = 1 - B(5;10,p')$.
The test has no ability to detect a false null hypothesis when $p > .90$ (see the
graph for .90 < p' < 1). This is a by-product of the unavoidable one-sided
rejection region (see **a** and **b**). The test also has an undesirably high $\beta(p')$ for
medium-to-large p', a result of the small sample size.

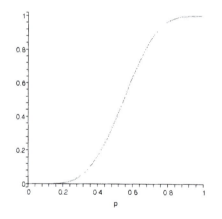

Chapter 10: Inferences Based on Two Samples

1.

    a. $E(\overline{X} - \overline{Y}) = E(\overline{X}) - E(\overline{Y}) = 4.1 - 4.5 = -.4$, irrespective of sample sizes.

    b. $V(\overline{X} - \overline{Y}) = V(\overline{X}) + V(\overline{Y}) = \dfrac{\sigma_1^2}{m} + \dfrac{\sigma_2^2}{n} = \dfrac{(1.8)^2}{100} + \dfrac{(2.0)^2}{100} = .0724$, and the s.d. of

    $\overline{X} - \overline{Y} = \sqrt{.0724} = .2691$.

    c. A normal curve with mean and s.d. as given in **a** and **b** (because m = n = 100, the CLT implies that both $\overline{X}$ and $\overline{Y}$ have approximately normal distributions, so $\overline{X} - \overline{Y}$ does also). The shape is not necessarily that of a normal curve when m = n = 10, because the CLT cannot be invoked. So if the two lifetime population distributions are not normal, the distribution of $\overline{X} - \overline{Y}$ will typically be quite complicated.

3.   The test statistic value is $z = \dfrac{(\overline{x} - \overline{y}) - 5000}{\sqrt{\dfrac{s_1^2}{m} + \dfrac{s_2^2}{n}}}$, and H_o will be rejected at level .01 if

    $z \geq 2.33$. We compute $z = \dfrac{(42{,}500 - 36{,}800) - 5000}{\sqrt{\dfrac{2200^2}{45} + \dfrac{1500^2}{45}}} = \dfrac{700}{396.93} = 1.76$, which is not >

2.33, so we don't reject H_o and conclude that the true average life for radials does not exceed that for economy brand by significantly more than 500.

5.

    a. H_a says that the average calorie output for sufferers is more than 1 cal/cm^2/min

    below that for non-sufferers. $\sqrt{\dfrac{\sigma_1^2}{m} + \dfrac{\sigma_2^2}{n}} = \sqrt{\dfrac{(.2)^2}{10} + \dfrac{(.4)^2}{10}} = .1414$, so

    $z = \dfrac{(.64 - 2.05) - (-1)}{.1414} = -2.90$. At level .01, H_o is rejected if $z \leq -2.33$; since –

2.90 < -2.33, reject H_o.

    b. $P = \Phi(-2.90) = .0019$

    c. $\beta = 1 - \Phi\left(-2.33 - \dfrac{-1.2 + 1}{.1414}\right) = 1 - \Phi(-.92) = .8212$

    d. $m = n = \dfrac{.2(2.33 + 1.28)^2}{(-.2)^2} = 65.15$, so use 66.

Chapter 10: Inferences Based on Two Samples

7.

1.   Parameter of interest: $\mu_1 - \mu_2 =$ the true difference of means for males and females on the Boredom Proneness Rating. Let $\mu_1 =$ men's average and $\mu_2 =$ women's average.

2. $H_o: \mu_1 - \mu_2 = 0$

3. $H_a: \mu_1 - \mu_2 > 0$

4. $z = \dfrac{(\bar{x} - \bar{y}) - \Delta_o}{\sqrt{\dfrac{s_1^2}{m} + \dfrac{s_2^2}{n}}} = \dfrac{(\bar{x} - \bar{y}) - 0}{\sqrt{\dfrac{s_1^2}{m} + \dfrac{s_2^2}{n}}}$

5.   RR: $z \geq 1.645$

6. $z = \dfrac{(10.40 - 9.26) - 0}{\sqrt{\dfrac{4.83^2}{97} + \dfrac{4.68^2}{148}}} = 1.83$

7.   Reject H_o. The data indicates the average Boredom Proneness Rating is higher for males than for females.

9.

a. point estimate $\bar{x} - \bar{y} = 19.9 - 13.7 = 6.2$. It appears that there could be a difference.

b.

$H_o: \mu_1 - \mu_2 = 0$, $H_a: \mu_1 - \mu_2 \neq 0$, $z = \dfrac{(19.9 - 13.7)}{\sqrt{\dfrac{39.1^2}{60} + \dfrac{15.8^2}{60}}} = \dfrac{6.2}{5.44} = 1.14$, and

the p-value $= 2[P(z > 1.14)] = 2(.1271) = .2542$. The p value is larger than any reasonable α, so we do not reject H_0. There is no significant difference.

c. No. With a normal distribution, we would expect most of the data to be within 2 standard deviations of the mean, and the distribution should be symmetric. 2 sd's above the mean is 98.1, but the distribution stops at zero on the left. The distribution is positively skewed.

d. We will calculate a 95% confidence interval for μ, the true average length of stays for patients given the treatment. $19.9 \pm 1.96 \dfrac{39.1}{\sqrt{60}} = 19.9 \pm 9.9 = (10.0, 21.8)$

Chapter 10: Inferences Based on Two Samples

11. $(\overline{X} - \overline{Y}) \pm z_{\alpha/2} \sqrt{\dfrac{s_1^2}{m} + \dfrac{s_2^2}{n}}$. Standard error $= \dfrac{s}{\sqrt{n}}$. Substitution yields

$(\overline{x} - \overline{y}) \pm z_{\alpha/2} \sqrt{(SE_1)^2 + (SE_2)^2}$. Using $\alpha = .05$, $z_{\alpha/2} = 1.96$, so

$(5.5 - 3.8) \pm 1.96 \sqrt{(0.3)^2 + (0.2)^2} = (0.99, 2.41)$. We are 95% confident that the true average blood lead level for male workers is between 0.99 and 2.41 higher than the corresponding average for female workers.

13. $\sigma_1 = \sigma_2 = .2$, $d = .2$, $\alpha = \beta = .05$, and the test is one-tailed, so

$n = \dfrac{(.04 + .04)(1.645 + 1.645)^2}{.04} = 21.65$, so use $n = 22$ hospitals of each type. We cannot make cause-and-effect conclusions here, since this is merely an observational study (nurse staffing problems were not forcibly introduced into randomly selected hospitals!). The general financial state of a hospital may impact both its nursing staff and its mortality rate.

15.

a. As either m or n increases, σ decreases, so $\dfrac{\mu_1 - \mu_2 - \Delta_o}{\sigma}$ increases (the numerator is positive), so $\left(z_\alpha - \dfrac{\mu_1 - \mu_2 - \Delta_o}{\sigma} \right)$ decreases, so

$\beta = \Phi \left(z_\alpha - \dfrac{\mu_1 - \mu_2 - \Delta_o}{\sigma} \right)$ decreases.

b. As β decreases, z_β increases, and since z_β is the numerator of n , n increases also.

17. Let μ_1 and μ_2 be the average time per week spent thinking about new ideas by all U.S. managers and all Canadian managers. We wish to test $H_0: \mu_1 = \mu_2$ versus $H_a: \mu_1 \neq \mu_2$. With such large sample sizes, construct a z-statistic: $z = \dfrac{(5.8 - 5.1) - 0}{\sqrt{\dfrac{6.0^2}{174} + \dfrac{4.6^2}{353}}} = 1.36$. The

two-sided P-value is $2(.0869) = .1738$, and so we fail to reject H_0 at any reasonable level of significance. The data do not suggest that the true average time per week that U.S. managers spend thinking about new ideas differs significantly from that for Canadian managers.

Chapter 10: Inferences Based on Two Samples

19. Let μ_1 and μ_2 be the average number of credit cards for <u>all</u> such "no-involvement" students and <u>all</u> such "payment-help" students, respectively. We wish to test H_0: $\mu_1 = \mu_2$ versus H_a: $\mu_1 > \mu_2$. With such large sample sizes, construct a z-statistic: $z = \dfrac{(2.22 - 2.09) - 0}{\sqrt{\dfrac{1.58^2}{209} + \dfrac{1.65^2}{75}}} = 0.59$. The P-value is $P(Z \geq 0.59) = 1 - \Phi(0.59) = .2776$, and so we fail to reject H_0 at any reasonable level of significance. The data do not suggest that students without parental involvement have a higher average number of credit cards than those with parental assistance.

21. With sample 1 begin amateurs and sample 2 being professionals, we wish to test the hypotheses H_o: $\mu_1 - \mu_2 = 0$ vs. H_a: $\mu_1 - \mu_2 < 0$. Calculating df as in the text gives $v = 42$, and the test statistic is $t = \dfrac{74.5 - 81.8}{\sqrt{\dfrac{6.29^2}{24} + \dfrac{8.64^2}{24}}} = -3.35$ The one-sided P-value is $P(t \leq -3.35) \approx P(t \geq 3.4) \approx .001$, using the df = 40 column of table A.8. So we reject H_o and conclude that, on average, expert pianists hit the keys harder than amateur pianists.

23.

a. Normal plots

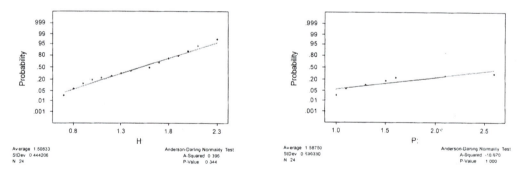

Using Minitab to generate normal probability plots, we see that both plots illustrate sufficient linearity. Therefore, it is plausible that both samples have been selected from normal population distributions.

b.

Comparative Box Plot for High Quality and Poor Quality Fabric

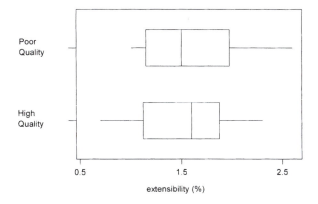

The comparative boxplot does not suggest a difference between average extensibility for the two types of fabrics.

c. We test $H_0 : \mu_1 - \mu_2 = 0$ vs. $H_a : \mu_1 - \mu_2 \neq 0$. With degrees of freedom

$$v = \frac{(.0433265)^2}{.00017906} = 10.5,$$ which we round down to 10, and using significance

level .05 (not specified in the problem), we reject H_o if $|t| \geq t_{.025,10} = 2.228$. The

test statistic is $t = \dfrac{-.08}{\sqrt{(.0433265)}} = -.38$, which is not ≥ 2.228 in absolute value,

so we cannot reject H_o. There is insufficient evidence to claim that the true average extensibility differs for the two types of fabrics.

25. Let μ_1 = the true average potential drop for alloy connections and let μ_2 = the true average potential drop for EC connections. Since we are interested in whether the potential drop is higher for alloy connections, an upper tailed test is appropriate. We test $H_0 : \mu_1 - \mu_2 = 0$ vs. $H_a : \mu_1 - \mu_2 > 0$. Using the SAS output provided, the test statistic, when assuming unequal variances, is t = 3.6362, the corresponding df is 37.5, and the p-value for our upper tailed test would be ½ (two-tailed p-value) = $\frac{1}{2}(.0008) = .0004$. Our p-value of .0004 is less than the significance level of .01, so we reject H_o. We have sufficient evidence to claim that the true average potential drop for alloy connections is higher than that for EC connections.

Chapter 10: Inferences Based on Two Samples

27. We will test the hypotheses: $H_0 : \mu_1 - \mu_2 = 10$ vs. $H_a : \mu_1 - \mu_2 > 10$. The test statistic is $t = \dfrac{(\bar{x} - \bar{y}) - 10}{\sqrt{\left(\frac{2.75^2}{10} + \frac{4.44^2}{5}\right)}} = \dfrac{4.5}{2.17} = 2.08$ The degrees of freedom

$$v = \dfrac{\left(\frac{2.75^2}{10} + \frac{4.44^2}{5}\right)^2}{\frac{\left(\frac{2.75^2}{10}\right)^2}{9} + \frac{\left(\frac{4.44^2}{5}\right)^2}{4}} = \dfrac{22.08}{3.95} = 5.59 \approx 5$$, and the p-value from table A.8 is approx

.045, which is $< .10$ so we reject H_0 and conclude that the true average lean angle for older females is more than 10 degrees smaller than that of younger females.

29.

a. Probability plots for the Coke and Pepsi data appear below. Both are fairly linear, supporting the requisite normality assumption.

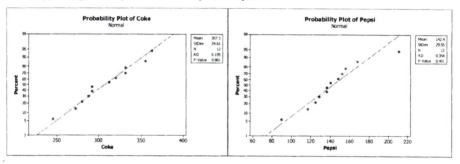

b. The mean and sd for the Coke data are 307.28 and 34.61, while the mean and sd for the Pepsi data are 142.44 and 29.55. The estimated degrees of freedom are $v = 21$, and the t critical value is $t_{.005,21} = 2.831$. The resulting 99% CI for the difference in population means is $(127.63, 202.03)$.

c. No, for a 99% lower confidence bound we use (difference of means) $- t_{.01,21}(\text{se}) = 131.75$.

d. We are 99% confident that the average foam volume from a 12 oz can of Coke is at least 131.75ml greater than the average foam volume from a 12 oz can of Pepsi.

Chapter 10: Inferences Based on Two Samples

31. Let μ_1 = the true average proportional stress limit for red oak and let μ_2 = the true average proportional stress limit for Douglas fir. We test $H_0 : \mu_1 - \mu_2 = 1$ vs. $H_a : \mu_1 - \mu_2 > 1$. The test statistic is $t = \dfrac{(8.48 - 6.65) - 1}{\sqrt{\frac{.79^2}{14} + \frac{1.28^2}{10}}} = \dfrac{1.83}{\sqrt{.2084}} 1.818$. With

degrees of freedom $\nu = \dfrac{(.2084)^2}{\frac{\left(\frac{.79^2}{14}\right)^2}{13} + \frac{\left(\frac{1.28^2}{10}\right)^2}{9}} = 13.85 \approx 13$, the p-value = $P(t > 1.8) =$

.048. We would reject H_0 at significance levels greater than .046 (e.g., the standard 5% significance level). At $\alpha = .05$, there is sufficient evidence to claim that true average proportional stress limit for red oak exceeds that of Douglas fir by more than 1 MPa.

33.

a. Following the usual format for most confidence intervals: *statistic $\pm$ (critical value)(standard error)*, a pooled variance confidence interval for the difference between two means is $(\bar{x} - \bar{y}) \pm t_{\alpha/2, m+n-2} \cdot s_p \sqrt{\frac{1}{m} + \frac{1}{n}}$.

b. The sample means and standard deviations of the two samples are $\bar{x} = 13.90$, $s_1 = 1.225$, $\bar{y} = 12.20$, $s_2 = 1.010$. The pooled variance estimate is $s_p^2 = \left(\frac{m-1}{m+n-2}\right)s_1^2 + \left(\frac{n-1}{m+n-2}\right)s_2^2 = \left(\frac{4-1}{4+4-2}\right)(1.225)^2 + \left(\frac{4-1}{4+4-2}\right)(1.010)^2$

$= 1.260$, so $s_p = 1.1227$. With df = m+n-1 = 6 for this interval,

$t_{.025,6} = 2.447$ and the desired interval is

$(13.90 - 12.20) \pm (2.447)(1.1227)\sqrt{\frac{1}{4} + \frac{1}{4}} = 1.7 \pm 1.943 = (-.24, 3.64)$. This interval contains 0, so it does not support the conclusion that the two population means are different.

c. Using the two-sample t interval discussed earlier, we use the CI as follows: First, we need to calculate the degrees of freedom.

$\nu = \dfrac{\left(\frac{1.225^2}{4} + \frac{1.01^2}{4}\right)^2}{\frac{\left(\frac{1.225^2}{4}\right)^2}{3} + \frac{\left(\frac{1.01^2}{4}\right)^2}{3}} = \dfrac{.3971}{.0686} = 5.78 \downarrow 5$ so $t_{.025,5} = 2.571$. Then the

interval is $(13.9 - 12.2) \pm 2.571\sqrt{\frac{1.225^2}{4} + \frac{1.01^2}{4}} = 1.70 \pm 2.571(.7938) = (-.34, 3.74)$. This interval is slightly wider, but it still supports the same conclusion.

Chapter 10: Inferences Based on Two Samples

35.

  a. It appears that bartenders pour slightly less rum into highball glasses, on average. But the most stark difference is variability: the amount poured into a slender, highball glass is much more consistent across bartenders than the amount poured into short, tumbler glasses. Both boxplots support an assumption of normally distributed populations.

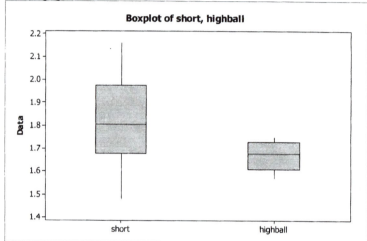

  b. As noted above, the two samples appear normal; probability plots confirm this. MINITAB reports the following: $t = 1.88$ with estimate df $= 8$. The corresponding two-sided P-value from MINITAB is 0.097; hence, we fail to reject the null hypothesis at the standard $\alpha = 0.05$ level. We conclude that the true average amount of rum poured by experienced bartenders does not differ significantly from tumblers to highball glasses.

37. Let μ_1 and μ_2 be the average OCSD scores for the appropriate populations of males and females, respectively. We wish to test H_0: $\mu_1 = \mu_2$ versus H_a: $\mu_1 \neq \mu_2$. The samples are moderate in size, so, we use a two-sample t test. MINITAB gives the following results: $t = 2.19$, estimated df $= 81$, P-value $= .031$. Hence, we reject the null hypothesis at the standard $\alpha = .05$. At this level, we conclude that the average OCSD scores are different for the populations of males and females with comorbid alcohol addition and PTSD. If we use the stricter $\alpha = .01$ standard instead, we would fail to reject H_0, because $.031 > .01$.

Chapter 10: Inferences Based on Two Samples

39.

a. $\bar{d} = 7.25$, $s_D = 11.8628$

1   Parameter of Interest: μ_D = true average difference of breaking load for fabric in unabraded or abraded condition.

2      $H_0 : \mu_D = 0$

3      $H_a : \mu_D > 0$

4      $t = \dfrac{\bar{d} - \mu_D}{s_D / \sqrt{n}} = \dfrac{\bar{d} - 0}{s_D / \sqrt{n}}$

5   RR: $t \geq t_{.01,7} = 2.998$

6      $t = \dfrac{7.25 - 0}{11.8628 / \sqrt{8}} = 1.73$

7   Fail to reject H_0. The data does not indicate a significant mean difference in breaking load for the two fabric load conditions.

b. No: a normal probability plot of the differences shows that difference #2 (55.0 – 20.0 = 35.0) is an extreme outlier, which casts doubt on the relevant normality assumption.

c. With fabric #2 deleted, the remaining observations create a reasonably linear normal probability plot. The revised values are $\bar{d} = 3.286$ $s_D = 4.183$, $t = 2.08 <$ $t_{.01,6} = 3.143$, and so we again fail to reject H_0. The data still does not indicate a significant mean difference in breaking load for the two fabric load conditions.

41.

a. We need to check the normality of the ten <u>differences</u>, so compute these first. A normal probability plot of those 10 differences appears below; the population of differences is plausibly normal.

167

Chapter 10: Inferences Based on Two Samples

b. Let μ_d denote the population mean difference (order = drug – no drug). From MINITAB, the mean and sd of the differences are 1.80 and 1.92, respectively. The t critical value is $t_{.025,9} = 2.262$, and the resulting 95% CI for μ_d is (0.426,3.176). The drugs seem to offer some help: on average, patients sleep between .426 hours and 3.176 hours longer with the drugs than without them.

43. We test $H_0 : \mu_d = 5$ vs. $H_a : \mu_d > 5$. With $\bar{d} = 7.600$, and $s_d = 4.178$,

$$t = \frac{7.600 - 5}{4.178/\sqrt{9}} = \frac{2.6}{1.39} = 1.87 \approx 1.9 .$$ With degrees of freedom $n - 1 = 8$, the corresponding p-value is P(t > 1.9) = .047. We would reject H_0 at any alpha level greater than .047. So, at the typical significance level of .05, we would reject H_0, and conclude that the data indicates that the higher level of illumination yields a decrease of more than 5 seconds in true average task completion time.

45.

a. Although there is a "jump" in the middle of the Normal Probability plot, the data follow a reasonably straight path, so there is no strong reason for doubting the normality of the population of differences.

b. A 95% lower confidence bound for the population mean difference is:

$$\bar{d} - t_{.05,14}\left(\frac{s_d}{\sqrt{n}}\right) = -38.60 - (1.761)\left(\frac{23.18}{\sqrt{15}}\right) = -38.60 - 10.54 = -49.14 .$$ We are 95% confident that the true mean difference between age at onset of Cushing's disease symptoms and age at diagnosis is greater than -49.14.

c. A 95% upper confidence bound for the corresponding population mean difference is 38.60 + 10.54 = 49.14.

47. With $(x_1, y_1) = (6,5)$, $(x_2, y_2) = (15,14)$, $(x_3, y_3) = (1,0)$, and $(x_4, y_4) = (21,20)$, $\bar{d} = 1$ and $s_d = 0$ (the d_l's are 1, 1, 1, and 1), while $s_1 = s_2 = 8.96$, so $s_p = 8.96$ and t = .16.

Chapter 10: Inferences Based on Two Samples

49.

a. H_o will be rejected if $|z| \geq 1.96$. With $\hat{p}_1 = \dfrac{63}{300} = .2100$, and

$$\hat{p}_2 = \frac{75}{180} = .4167, \ \hat{p} = \frac{63+75}{300+180} = .2875,$$

$$z = \frac{.2100 - .4167}{\sqrt{(.2875)(.7125)\left(\frac{1}{300} + \frac{1}{180}\right)}} = \frac{-.2067}{.0427} = -4.84. \ \text{Since} -4.84 \leq -1.96, \ H_o$$

is rejected.

b. $\bar{p} = .275$ and $\hat{p}_1 = .150$, so power =

$$1 - \left[\Phi\left(\frac{[(1.96)(.0421) + .2]}{.0432}\right) - \Phi\left(\frac{[-(1.96)(.0421) + .2]}{.0432}\right)\right] =$$
$$1 - \left[\Phi(6.54) - \Phi(2.72)\right] = .9967.$$

51. Let $\alpha = .05$. A 95% confidence interval is $(\hat{p}_1 - \hat{p}_2) \pm z_{\alpha/2} \sqrt{\left(\frac{\hat{p}_1 \hat{q}_1}{m} + \frac{\hat{p}_2 \hat{q}_2}{n}\right)}$

$$= \left(\tfrac{224}{395} - \tfrac{126}{266}\right) \pm 1.96 \sqrt{\left(\frac{\left(\frac{224}{395}\right)\left(\frac{171}{395}\right)}{395} + \frac{\left(\frac{126}{266}\right)\left(\frac{140}{266}\right)}{266}\right)} = .0934 \pm .0774 = (.0160, .1708).$$

53. Let p_1 = true proportion of irradiated bulbs that are marketable; p_2 = true proportion of untreated bulbs that are marketable; The hypotheses are $H_0 : p_1 - p_2 = 0$ vs.
$H_0 : p_1 - p_2 > 0$. The test statistic is $z = \dfrac{\hat{p}_1 - \hat{p}_2}{\sqrt{\hat{p}\hat{q}\left(\frac{1}{m} + \frac{1}{n}\right)}}$. With $\hat{p}_1 = \dfrac{153}{180} = .850$,

and $\hat{p}_2 = \dfrac{119}{180} = .661$, $\hat{p} = \dfrac{272}{360} = .756$, $z = \dfrac{.850 - .661}{\sqrt{(.756)(.244)\left(\frac{1}{180} + \frac{1}{180}\right)}} = \dfrac{.189}{.045} = 4.2$.

The p-value = $1 - \Phi(4.2) \approx 0$, so reject H_o at any reasonable level. Radiation appears to be beneficial.

Chapter 10: Inferences Based on Two Samples

55.

a. The "after" success probability is $p_1 + p_3$ while the "before" probability is $p_1 + p_2$, so $p_1 + p_3 > p_1 + p_2$ becomes $p_3 > p_2$; thus we wish to test $H_0 : p_3 = p_2$ versus $H_a : p_3 > p_2$.

b. The estimator of $(p_1 + p_3) - (p_1 + p_2)$ is
$$\frac{(X_1 + X_3) - (X_1 + X_2)}{n} = \frac{X_3 - X_2}{n}.$$

c. When H_o is true, $p_2 = p_3$, so $Var\left(\dfrac{X_3 - X_2}{n}\right) = \dfrac{p_2 + p_3}{n}$, which is estimated by $\dfrac{\hat{p}_2 + \hat{p}_3}{n}$. The Z statistic is then
$$\frac{\dfrac{X_3 - X_2}{n}}{\sqrt{\dfrac{\hat{p}_2 + \hat{p}_3}{n}}} = \frac{X_3 - X_2}{\sqrt{X_2 + X_3}}.$$

d. The computed value of Z is $\dfrac{200 - 150}{\sqrt{200 + 150}} = 2.68$, so
$P = 1 - \Phi(2.68) = .0037$. At level .01, H_o can be rejected but at level .001 H_o would not be rejected.

57. Using $p_1 = q_1 = p_2 = q_2 = .5$, $L = 2(1.96)\sqrt{\dfrac{.25}{n} + \dfrac{.25}{n}} = \dfrac{2.7719}{\sqrt{n}}$, so L=.1 requires
n=769.

59. Let p_1 and p_2 stand for the proportions of all left-handed and right-handed male Navy enlisted personnel who have been hospitalized for injuries. We wish to test H_0: $p_1 = p_2$ versus H_a: $p_1 \ne p_2$. The large-sample z test statistic is $z = \dfrac{(\hat{p}_1 - \hat{p}_2) - 0}{\sqrt{\overline{pq}\left(\dfrac{1}{n_1} + \dfrac{1}{n_2}\right)}} = \ldots =$

3.14, with a corresponding P-value of $P(|Z| \ge 3.14) = .002$. Hence, we reject the null hypothesis at any reasonable significance level; proneness to injury seems to differ significantly between these two populations (the point estimates are ~39% for left-handers and ~29% for right-handers). However, as with any observational study, no causal connection can be drawn!

Chapter 10: Inferences Based on Two Samples

61.

    a. Since the given f value of 4.75 falls between $F_{.05,5,10} = 3.33$ and

        $F_{.01,5,10} = 5.64$, we can say that the upper-tailed p-value is between .01 and .05.

    b. Since the given f of 2.00 is less than $F_{.10,5,10} = 2.52$, the p-value > .10.

    c. The two tailed p-value = $2P(F \geq 5.64) = 2(.01) = .02$.

    d. For a lower tailed test, we must first use formula 9.9 to find the critical values:

$$F_{.90,5,10} = \frac{1}{F_{.10,10,5}} = .3030, \quad F_{.95,5,10} = \frac{1}{F_{.05,10,5}} = .2110,$$

$$F_{.99,5,10} = \frac{1}{F_{.01,10,5}} = .0995. \text{ Since } .0995 < f = .200 < .2110, \quad .01 < \text{p-value} < .05$$

        (but obviously closer to .05).

    e. There is no column for numerator d.f. of 35 in Table A.9, however looking at both df = 30 and df = 40 columns, we see that for denominator df = 20, our f value is between $F_{.01}$ and $F_{.001}$. So we can say .001< p-value <.01.

63. With σ_1 = true standard deviation for not-fused specimens and σ_2 = true standard deviation for fused specimens, we test $H_0 : \sigma_1 = \sigma_2$ vs. $H_a : \sigma_1 > \sigma_2$. The calculated test statistic is $f = \dfrac{(277.3)^2}{(205.9)^2} = 1.814$. With numerator d.f. = m – 1 = 10 – 1 = 9, and denominator d.f. = n – 1 = 8 – 1 = 7, $f = 1.814 < 2.72 = F_{.10,9,7}$. We can say that the p-value > .10, which is obviously > .01, so we cannot reject H_0. There is not sufficient evidence that the standard deviation of the strength distribution for fused specimens is smaller than that of not-fused specimens.

65. For the hypotheses $H_0 : \sigma_1 = \sigma_2$ versus $H_a : \sigma_1 \neq \sigma_2$, we find a test statistic of f = 1.22. At df = (47,44) ≈ (40,40), 1.22 < 1.51 indicates the P-value is greater than 2(.10) = .20. Hence, H_0 is not rejected. The data does not suggest a significant difference in the two population variances.

Chapter 10: Inferences Based on Two Samples

67. A 95% upper bound for $\dfrac{\sigma_2}{\sigma_1}$ is $\sqrt{\dfrac{s_2^2 F_{.05,9,9}}{s_1^2}} = \sqrt{\dfrac{(3.59)^2 (3.18)}{(.79)^2}} = 8.10$. We are

confident that the ratio of the standard deviation of triacetate porosity distribution to that of the cotton porosity distribution is at most 8.10.

69.

    a. Minitab gives the following results: group L has a mean gpa of 3.367 with a sd of 0.514; group N has a mean gpa of 2.920 with a sd of 0.598; the estimated df is v = 56. From these, a 95% CI for $\mu_1 - \mu_2$ is (0.158,.735).

    b. You can create the bootstrap distribution of differences by using the macro from Chapter 8 separately on each of the two samples, then computing differences of the side-by-side pairs. Answers will vary, but the bootstrap distribution of differences looks quite normal.

    c. Answers will vary; one simulation gave $s_{boot} = 0.141$. This suggests the following 95% CI for $\mu_1 - \mu_2$: $(3.367 - 2.920) \pm t_{.025,56}(0.141) \approx (0.447) \pm (2)(0.141) = (.165,.729)$.

    d. Answers will vary; choosing the 25th bootstrap value from each end of the distribution in one simulation gave a percentile interval of (.156,.740).

    e. All three intervals are very close to each other, suggesting the sampling distribution of the difference of means is normal here, as noted above in (b).

    f. Students on lifestyle floors appear to have a higher mean gpa, somewhere between ~.16 higher and ~.73 higher.

71.

    a. You can create the bootstrap distribution of differences by using the median macro from Chapter 8 separately on each of the two samples, then computing differences of the side-by-side pairs. The bootstrap distribution of differences of medians is definitely not normal: the distribution is multimodal and positively skewed.

    b. Answers will vary; one simulation gave $s_{boot} = 2.5657$. The medians of the two original samples are 13.88 and 8.47. This suggests the following 95% CI for $\tilde{\mu}_1 - \tilde{\mu}_2$: $(13.88 - 8.47) \pm t_{.025,16}(2.5657) \approx (5.41) \pm (2.120)(2.5657) = (-0.029,10.85)$.

    c. Answers will vary; choosing the 25th bootstrap value from each end of the distribution in one simulation gave a percentile interval of (0.4706,10.0294).

172

d. The interval in (c) is narrower and, in particular, does not include zero. We have more faith in (c), since (b) relied on a normally distributed sampling distribution, which does not exist here.

e. The intervals from Exercise 70 are considerably narrower (more "precise") than those for the difference in population medians. We can more precisely measure the difference in population means with the bootstrap in this particular case.

73.

a. The standard deviations of the two samples are 0.514 and 0.598. The relevant critical value is $F_{.025,29,29} = 2.101$ (from Minitab). Thus, a 95% CI for $\dfrac{\sigma_1}{\sigma_2}$ is

$$\left(\frac{0.514}{0.598} \frac{1}{\sqrt{2.101}}, \frac{0.514}{0.598} \sqrt{2.101} \right) = (0.593, 1.246).$$ Normal probability plots of the two samples shows some noticeable departures from normality, more so that we are usually willing to accept for this F procedure.

b. Use the macro below separately on each sample to create parallel columns of 1000 bootstrap standard deviations from each sample. Then calculate the 1000 ratios.
```
Let k3 = N(c1)
sample k3 c1 c3;
replace.
let k1=stdev(c3)
stack k1 c5 c5
end
```
Find the 25th value from either end of the distribution of ratios. Answers will vary; one bootstrap simulation gave (0.696, 1.472).

c. The interval in (b) is narrower. We have greater faith in this interval, too, since (a) was based upon a dubious normality assumption. Notice that 1 appears in these intervals, suggesting the two population sd's could be equal. This is consistent with Exercise 72.

75.

a. For the test of H_0: $\mu_1 - \mu_2 = 0$ versus H_a: $\mu_1 - \mu_2 \neq 0$, our test statistic is $t =$

$$\frac{(10.59 - 5.71) - 0}{\sqrt{\dfrac{4.41^2}{10} + \dfrac{3.92^2}{10}}} = 2.62;$$ from Minitab, the estimated df is $v = 17$. The 2-sided

P-value is roughly $2P(|t| > 2.6) = 2(.009) = .018$. Hence, we reject H_0 at the $\alpha =$

Chapter 10: Inferences Based on Two Samples

.05 level and conclude the two population means are different. Neither of the probability plots looks very linear, but it's difficult to detect moderate deviations from normality with so few observations.

b. Stack the two samples in C1, with their corresponding subscripts (here, C and B) in C2. The following macro will randomly permute the subscripts, create samples accordingly, and compute the ratio of the standard deviations of the permuted samples. The resulting values are stacked in C10.

```
Let k3 = N(c1)
sample k3 c2 c3
unstack c1 c4 c5;
subs c3.
let k1=mean(c4)-mean(c5)
stack k1 c10 c10
end
```

Run this macro 999 times. Then find the proportion of these differences in means that are greater than our observed difference, $10.59 - 5.71 = 4.88$. Double this proportion to get the 2-sided P-value. Answers will vary; in one bootstrap simulation, only 11 of the 999 simulated differences were above 4.88, giving $2[12/1000]$ .024 as our 2-sided P-value

c. The answers to (a) and (b) are quite similar; in particular, both reject the null hypothesis of equal means at the $\alpha = .05$ level. This is not surprising, since the sampling distribution relevant to (a) was indeed normal (see Exercise 74).

77.

a. The standard deviations of the two samples are 3.26 and 1.54, for a F-ratio of $f = 4.46$. Compare this to $F_{.05,6,5} = 4.95$ and $F_{.95,6,5} = 1/F_{.05,5,6} = 1/4.39 = 0.228$: since $0.228 < 4.46 < 4.95$, we fail to reject the hypothesis that $\sigma_1 = \sigma_2$ at the $\alpha = .10$ level. The finaska barley group shows some deviation from normality, but it's difficult to detect a real departure with such a small sample.

b. Stack the two samples in C1, with their corresponding subscripts (here, T and F) in C2. The following macro will randomly permute the subscripts, create samples accordingly, and compute the ratio of the standard deviations of the permuted samples. The resulting values are stacked in C10.

```
Let k3 = N(c1)
sample k3 c2 c3
unstack c1 c4 c5;
subs c3.
let k1=stdev(c4)/stdev(c5)
stack k1 c10 c10
end
```

Chapter 10: Inferences Based on Two Samples

Run this macro 999 times. Then find the proportion of these ratios that are more than our observed sd ratio, $3.26/1.54 = 2.117$. Double this proportion to get the 2-sided P-value. Answers will vary; in one bootstrap simulation, 85 of the 999 simulated ratios were above 2.117, giving $2[86/1000] = .172$ as our 2-sided P-value.

    c.   In either case, we have no statistically significant evidence to suggest the population standard deviations are unequal.

79.

    a.   Use the macro provided in Exercise 75(b). In one run of 999 permutations, 18 of the differences in means were less than the observed difference, -3.47. The resulting 2-sided P-value is $2[19/1000] = .038$. Thus, we reject the null hypothesis of equal population means at the $\alpha = .05$ level.

    b.   The result in (a) matches closely the result in Example 10.4; even the P-values are fairly close (.032 v .038). This comes as no surprise, since both procedures are valid: the large sample sizes permit a large-sample z-test, and the shapes of the distributions of the two samples are fairly similar (which is important for the validity of the permutation test).

81.

    a.   Minitab gives the following results: $\bar{d} = 9.126$, $s_d = 6.893$. So, a 95% confidence interval for μ_d is $9.126 \pm t_{.025,26}(6.893)/\sqrt{27} = (6.399, 11.853)$. The 27 differences are grossly non-normal (heavily left-skewed); however, with a moderate sample size of $n = 27$, the effects of the CLT may begin to appear in the sampling distribution of $\bar{d}$.

    b.   Use the macro provided in Chapter 8. The bootstrap distribution of $\bar{d}$ is still quite non-normal (left-skewed).

    c.   Answers will vary; one simulation gave $s_{boot} = 1.305$. This suggests the following 95% CI for μ_d: $(9.126) \pm t_{.025,26}(1.305) = (9.126) \pm (2.056)(1.305) = (6.44, 11.81)$.

    d.   Answers will vary; choosing the 25th bootstrap value from each end of the distribution in one simulation gave a percentile interval of $(6.23, 11.51)$.

    e.   The intervals in (a) and (c) are quite similar; however, the interval in (d) is clearly shifted to the left, reflecting the left-skewedness of the sampling/bootstrap distribution of $\bar{d}$. This calls into question the intervals of (a) and (c).

    f.   On average, books cost between $6.23 and $11.51 more with Amazon than at the campus bookstore. In light of this, we might stop buying textbooks on-line!

Chapter 10: Inferences Based on Two Samples

83. Both the bootstrap and the randomized permutation test simulate random sampling from a desired distribution, in order to provide a confidence interval (bootstrap only) or to test a hypothesis (either method). The bootstrap method assumes our sample faithfully represents its population, so that sampling with replacement from the sample is equivalent to creating iid observations from the population. We then use these bootstrap samples to create a faithful representation of the sampling distribution of our relevant statistic (a sample mean or sd, a difference of two means, a median, whatever). Permutation tests are only used for comparison of two populations, and we make a different assumption: under the null hypothesis, the two populations of interest are identically distributed, and so our $m+n$ observations are really from the same distribution.

85. With sample sizes 56 and 59, the degrees of freedom must be at least 55 (see Exercise 57; notice we cannot estimate v because we do not have the standard deviations). Thus, from Table A.5, $t = 6.07$ is statistically significant at any α level: .05, .01, .001. The mean number of ingredients selected by the scale-down group is indeed significantly greater than for the build-up group. This same principle might be applied to features on a new car, for example.

87. Since $m < n$, $v = \dfrac{[(se_1)^2 + (se_2)^2]^2}{\dfrac{(se_1)^4}{m-1} + \dfrac{(se_2)^4}{n-1}} > \dfrac{[(se_1)^2 + (se_2)^2]^2}{\dfrac{(se_1)^4}{m-1} + \dfrac{(se_2)^4}{m-1}}$

$= (m-1)\dfrac{[(se_1)^2 + (se_2)^2]^2}{(se_1)^4 + (se_2)^4}$; replacing n by m above increased the denominator,

which decreased the overall fraction. Then, if we expand the numerator of the

remaining fraction, $\dfrac{[(se_1)^2 + (se_2)^2]^2}{(se_1)^4 + (se_2)^4} = \dfrac{(se_1)^4 + (se_2)^4 + 2(se_1)^2(se_2)^2}{(se_1)^4 + (se_2)^4} > 1$, and

we conclude $v > (m-1)(1) = m-1$. So, a conservative estimate of the df for the 2-sample t procedures is $\min(m-1, n-1)$. This is easier to compute, but lowering df with result in a wider margin of error (for a CI) or less power (for a hypothesis test).

Chapter 10: Inferences Based on Two Samples

89.

a.

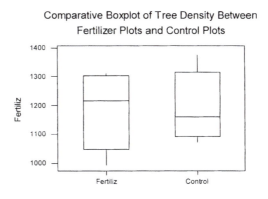

Comparative Boxplot of Tree Density Between Fertilizer Plots and Control Plots

Although the median of the fertilizer plot is higher than that of the control plots, the fertilizer plot data appears negatively skewed, while the opposite is true for the control plot data.

b. A test of $H_0 : \mu_1 - \mu_2 = 0$ vs. $H_a : \mu_1 - \mu_2 \neq 0$ yields a t value of -.20, and a two-tailed p-value of .85. (d.f. = 13). We would fail to reject H_0; the data does not indicate a significant difference in the means.

c. With 95% confidence we can say that the true average difference between the tree density of the fertilizer plots and that of the control plots is somewhere between – 144 and 120. Since this interval contains 0, 0 is a plausible value for the difference, which further supports the conclusion based on the p-value.

Chapter 10: Inferences Based on Two Samples

91. The center of any confidence interval for $\mu_1 - \mu_2$ is always $\bar{x}_1 - \bar{x}_2$, so

$$\bar{x}_1 - \bar{x}_2 = \frac{-473.3 + 1691.9}{2} = 609.3 \,.$$ Furthermore, half of the width of this interval is

$$\frac{1691.9 - (-473.3)}{2} = 1082.6 \,.$$ Equating this value to the expression on the right of the

95% confidence interval formula, $1082.6 = (1.96)\sqrt{\dfrac{s_1^2}{n_1} + \dfrac{s_2^2}{n_2}}$, we find

$$\sqrt{\frac{s_1^2}{n_1} + \frac{s_2^2}{n_2}} = \frac{1082.6}{1.96} = 552.35 \,.$$ For a 90% interval, the associated z value is 1.645, so

the 90% confidence interval is then $609.3 \pm (1.645)(552.35) = 609.3 \pm 908.6$
$= (-299.3, 1517.9)$.

93. $m = n = 40$, $\bar{x} = 3975.0$, $s_1 = 245.1$, $\bar{y} = 2795.0$, $s_2 = 293.7$. The large sample

99% confidence interval for $\mu_1 - \mu_2$ is $(3975.0 - 2795.0) \pm 2.58\sqrt{\dfrac{245.1^2}{40} + \dfrac{293.7^2}{40}}$

$(1180.0) \pm 1560 \approx (1020, 1340)$. The value 0 is not contained in this interval so
we can state that, with very high confidence, the value of $\mu_1 - \mu_2$ is not 0, which is
equivalent to concluding that the population means are not equal.

95. Let μ_1 denote the true average tear length for Brand A and let μ_2 denote the true
average tear length for Brand B. The relevant hypotheses are $H_0 : \mu_1 - \mu_2 = 0$ vs.
$H_a : \mu_1 - \mu_2 > 0$. Assuming both populations have normal distributions, the two-
sample t test is appropriate. m = 16, $\bar{x} = 74.0$, $s_1 = 14.8$, n = 14, $\bar{y} = 61.0$, $s_2 =$

12.5, so the approximate d.f. is $\nu = \dfrac{\left(\frac{14.8^2}{16} + \frac{12.5^2}{14}\right)^2}{\dfrac{\left(\frac{14.8^2}{16}\right)^2}{15} + \dfrac{\left(\frac{12.5^2}{14}\right)^2}{13}} = 27.97$, which we round down

to 27. The test statistic is $t = \dfrac{74.0 - 61.0}{\sqrt{\frac{14.8^2}{16} + \frac{12.5^2}{14}}} \approx 2.6$. From Table A.7, the p-value = P(

t > 2.6) = .007. At a significance level of .05, H_o is rejected and we conclude that the
average tear length for Brand A is larger than that of Brand B.

Chapter 10: Inferences Based on Two Samples

97. The relevant hypotheses are $H_0 : \mu_1^* - \mu_2^* = 0$ (which is equivalent to saying $\mu_1 - \mu_2 = 0$) versus $H_a : \mu_1^* - \mu_2^* \neq 0$ (which is the same as saying $\mu_1 - \mu_2 \neq 0$). The pooled t test is based on d.f. = m + n − 2 = 8 + 9 − 2 = 15. The pooled variance is $s_p^2 = \left(\dfrac{m-1}{m+n-2} \right) s_1^2 + \left(\dfrac{n-1}{m+n-2} \right) s_2^2$

$\left(\dfrac{8-1}{8+9-2} \right) (4.9)^2 + \left(\dfrac{9-1}{8+9-2} \right) (4.6)^2 = 22.49$, so $s_p = 4.742$. The test

statistic is $t = \dfrac{\overline{x}^* - \overline{y}^*}{s_p \sqrt{\frac{1}{m} + \frac{1}{n}}} = \dfrac{18.0 - 11.0}{4.742 \sqrt{\frac{1}{8} + \frac{1}{9}}} = 3.04 \approx 3.0$. From Table A.7, the

p-value associated with t = 3.0 is 2P(t > 3.0) = 2(.004) = .008. At significance level .05, H_o is rejected and we conclude that there is a difference between μ_1^* and μ_2^*, which is equivalent to saying that there is a difference between μ_1 and μ_2.

99. Let μ_1 denote the true average ratio for young men and μ_2 denote the true average ratio for elderly men. Assuming both populations from which these samples were taken are normally distributed, the relevant hypotheses are $H_0 : \mu_1 - \mu_2 = 0$ vs.

$H_a : \mu_1 - \mu_2 > 0$. The value of the test statistic is $t = \dfrac{(7.47 - 6.71)}{\sqrt{\dfrac{(.22)^2}{13} + \dfrac{(.28)^2}{12}}} = 7.5$.

The d.f. = 20 and the p-value is P(t > 7.5) ≈ 0. Since the p-value is $< \alpha = .05$, we reject H_o. We have sufficient evidence to claim that the true average ratio for young men exceeds that for elderly men.

101. NO, since a 2-sample t test is the wrong analysis here! Instead, we should perform a paired t test. For the data provided, $\overline{d} = 0.3$, $s_D = 0.276$, and t = 2.67 at 5df. This has a corresponding 2-sided P-value of 0.045, and so we reject the hypothesis of zero mean difference at the $\alpha = .05$ significance level.

Chapter 10: Inferences Based on Two Samples

103. Because of the nature of the data, we will use a paired t test. We obtain the differences by subtracting intake value from expenditure value. We are testing the hypotheses H_0: $\mu_d = 0$ vs H_a: $\mu_d \neq 0$. Test statistic $t = \dfrac{1.757}{1.197/\sqrt{7}} = 3.88$ with df $= n - 1 = 6$ leads to a p-value of $2[P(t>3.88)] \approx .008$. Using either significance level .05 or .01, we would reject the null hypothesis and conclude that there is a difference between average intake and expenditure. However, at significance level .001, we would not reject

105. Let p_1 = true survival rate at $11°C$; p_2 = true survival rate at $30°C$; The hypotheses are $H_0 : p_1 - p_2 = 0$ vs. $H_a : p_1 - p_2 \neq 0$. The test statistic is

$z = \dfrac{\hat{p}_1 - \hat{p}_2}{\sqrt{\hat{p}\hat{q}\left(\frac{1}{m} + \frac{1}{n}\right)}}$. With $\hat{p}_1 = \dfrac{73}{91} = .802$, and $\hat{p}_2 = \dfrac{102}{110} = .927$, $\hat{p} = \dfrac{175}{201} = .871$,

$\hat{q} = .129$. $z = \dfrac{.802 - .927}{\sqrt{(.871)(.129)\left(\frac{1}{91} + \frac{1}{110}\right)}} = \dfrac{-.125}{.0475} = -2.63$. The p-value =

$2\Phi(-2.63) = 2(.0043) = .0086$, so reject H_o at most reasonable levels (.10, .05, .01). The two survival rates appear to differ.

107. $\Delta_0 = 0$, $\sigma_1 = \sigma_2 = 10$, d = 1, $\sigma = \sqrt{\dfrac{200}{n}} = \dfrac{14.142}{\sqrt{n}}$, so $\beta = \Phi\left(1.645 - \dfrac{\sqrt{n}}{14.142}\right)$,

giving $\beta = .9015, .8264, .0294$, and .0000 for n = 25, 100, 2500, and 10,000 respectively. If the $\mu_i's$ referred to true average IQ's resulting from two different conditions, $\mu_1 - \mu_2 = 1$ would have little practical significance, yet very large sample sizes would yield statistical significance in this situation.

109. $H_0 : p_1 = p_2$ will be rejected at level α in favor of $H_a : p_1 > p_2$ if either $z \geq z_{.05} = 1.645$. With $\hat{p}_1 = \frac{250}{2500} = .10$, $\hat{p}_2 = \frac{167}{2500} = .0668$, and $\hat{p} = .0834$,

$z = \dfrac{.0332}{.0079} = 4.2$, so H_o is rejected . It appears that a response is more likely for a white name than for a black name.

111. First, $Var(\overline{X} - \overline{Y}) = \dfrac{\lambda_1}{m} + \dfrac{\lambda_2}{n} = \lambda\left(\dfrac{1}{m} + \dfrac{1}{n}\right)$ under H_0, where λ can be estimated

for the variance by the pooled estimate $\hat{\lambda} = \dfrac{m\overline{X} + n\overline{Y}}{m + n}$. With the obvious point

estimates $\hat{\lambda}_1 = \overline{X}$, $\hat{\lambda}_2 = \overline{Y}$, we have a large-sample test statistic of

$Z = \dfrac{(\overline{X} - \overline{Y}) - 0}{\sqrt{\hat{\lambda}\left(\dfrac{1}{m} + \dfrac{1}{n}\right)}} = \dfrac{\overline{X} - \overline{Y}}{\sqrt{\dfrac{\overline{X}}{n} + \dfrac{\overline{Y}}{m}}}$. With $\overline{x} = 1.616$ and $\overline{y} = 2.557$, $z = $ -5.3 and p-

value $= 2(\Phi(-5.3)) < .0006$, so we would certainly reject $H_0 : \lambda_1 = \lambda_2$ in favor

of $H_a : \lambda_1 \neq \lambda_2$

113.

a. We must show that, when H_0 is true, $P(R_1 \cap R_2) \leq \alpha$. Under the null hypothesis, either $\theta \in \Omega_1$ or $\theta \in \Omega_2$. Consider the first case: $P(R_1 \cap R_2$ when $\theta \in \Omega_1) \leq P(R_1$ when $\theta \in \Omega_1) = \alpha$, since we assume R_1 is a level α rejection region for $H_{01}: \theta \in \Omega_1$. Similarly, for the second case, $P(R_1 \cap R_2$ when $\theta \in \Omega_2) \leq P(R_2$ when $\theta \in \Omega_2) = \alpha$, since we assume R_2 is a level α rejection region for $H_{02}: \theta \in \Omega_2$.

b. At the $\alpha = .05$ level, the rejection region for an upper-tailed test with $v = 20$ is $t > t_{.05,20} = 1.725$, and the rejection region for a lower-tailed test is $t < -t_{.05,20} = -1.725$. Under the UIT, we reject H_0 in favor of bio-equivalence only if $t_L > 1.725$ and $t_U < -1.725$. In cases (i) and (ii), only one of these is satisfied, and we fail to reject the null hypothesis; in case (iii), both are satisfied and we can reject H_0 at the $\alpha = .05$ level.

Chapter 11: The Analysis of Variance

1.

a. H_o will be rejected if $f \geq F_{.05,4,15} = 3.06$ (since $I - 1 = 4$, and $I(J-1) = (5)(3)$ $= 15$). The computed value of F is $f = \dfrac{MSTr}{MSE} = \dfrac{2673.3}{1094.2} = 2.44$. Since 2.44 is not ≥ 3.06, H_o is not rejected. The data does not indicate a difference in the mean tensile strengths of the different types of copper wires.

b. $F_{.05,4,15} = 3.06$ and $F_{.10,4,15} = 2.36$, and our computed value of 2.44 is between those values, it can be said that $.05 < \text{p-value} < .10$.

3. $x_{\bullet\bullet} = IJ\bar{x}_{\bullet\bullet} = 32(5.19) = 166.08$, so $SST = 911.91 - \dfrac{(166.08)^2}{32} = 49.95$.

$SSTr = 8\left[(4.39 - 5.19)^2 + \ldots + (6.36 - 5.19)^2\right] = 20.38$, so

$SSE = 49.95 - 20.38 = 29.57$. Then $f = \dfrac{20.38/3}{29.57/28} = 6.43$. Since

$6.43 \geq F_{.05,3,28} = 2.95$, $H_0 : \mu_1 = \mu_2 = \mu_3 = \mu_4$ is rejected at level .05. There are differences between at least two average flight times for the four treatments.

5.

Source	Df	SS	MS	F
Treatments	3	509.112	169.707	10.85
Error	36	563.134	15.643	
Total	39	1,072.256		

$F_{.01,3,36} \approx F_{.01,3,30} = 4.51$. The computed test statistic value of 10.85 exceeds 4.51, so reject H_o in favor of H_a: at least two of the four means differ.

7. The summary quantities are $x_{1\bullet} = 2332.5$, $x_{2\bullet} = 2576.4$, $x_{3\bullet} = 2625.9$, $x_{4\bullet} = 2851.5$, $x_{5\bullet} = 3060.2$, $x_{\bullet\bullet} = 13,446.5$, so CF = 5,165,953.21, SST = 75,467.58, SSTr = 43,992.55, SSE = 31,475.03, $MSTr = \dfrac{43,992.55}{4} = 10,998.14$, $MSE = \dfrac{31,475.03}{30} = 1049.17$ and $f = \dfrac{10,998.14}{1049.17} = 10.48$. (These values should be displayed in an ANOVA table as requested.) Since $10.48 \geq F_{.01,4,30} = 4.02$, $H_0 : \mu_1 = \mu_2 = \mu_3 = \mu_4 = \mu_5$ is rejected. There are differences in the true average axial stiffness for the different plate lengths.

Chapter 11: The Analysis of Variance

9. $\sum\sum (x_{ij} - \bar{x}..)^2 = $ SST by definition. We want to show the sum of squares of the right-hand side of (11.2) is, in fact, SSTr + SSE. Expand the terms inside:

$$\sum\sum [(x_{ij} - \bar{x}_{i.}) + (\bar{x}_{i.} - \bar{x}..)]^2 = \sum\sum (x_{ij} - \bar{x}_{i.})^2 + 2\sum\sum (x_{ij} - \bar{x}_{i.})(\bar{x}_{i.} - \bar{x}..) +$$

$$\sum\sum (\bar{x}_{i.} - \bar{x}..)^2 = \text{SSE} + 2\sum\sum (x_{ij} - \bar{x}_{i.})(\bar{x}_{i.} - \bar{x}..) + \text{SSTr}.$$ Now, it remains to show the middle term is zero. To that end, watch the indexes on the sums carefully: $\sum_i \sum_j$
$(x_{ij} - \bar{x}_{i.})(\bar{x}_{i.} - \bar{x}..) = \sum_i [(\bar{x}_{i.} - \bar{x}..) \sum_j (x_{ij} - \bar{x}_{i.})] = \sum_i [(\bar{x}_{i.} - \bar{x}..)(0)] = \sum_i (0) = 0.$ For every j, the sum $\sum_j (x_{ij} - \bar{x}_{i.})$ equal zero because it's the sum of the (un-squared) deviations of the jth sample about its own sample mean.

11. $Q_{.05,5,15} = 4.37$, $w = 4.37\sqrt{\dfrac{272.8}{4}} = 36.09$.

3	1	4	2	5
437.5	462.0	469.3	512.8	532.1

The brands seem to divide into two groups: 1, 3, and 4; and 2 and 5; with no significant differences within each group but all between group differences are significant.

13.

3	1	4	2	5
427.5	462.0	469.3	502.8	532.1

Brand 1 does not differ significantly from 3 or 4, 2 does not differ significantly from 4 or 5, 3 does not differ significantly from 1, 4 does not differ significantly from 1 or 2, 5 does not differ significantly from 2, but all other differences (e.g., 1 with 2 and 5, 2 with 3, etc.) do appear to be significant.

Chapter 11: The Analysis of Variance

15. $Q_{.01,4,36} = 4.75$, $w = 4.75\sqrt{\dfrac{15.64}{10}} = 5.94$.

2	1	3	4
24.69	26.08	29.95	33.84

Treatment 4 appears to differ significantly from both 1 and 2, but there are no other significant differences.

17. $\theta = \Sigma c_i \mu_i$ where $c_1 = c_2 = .5$ and $c_3 = -1$, so

$\hat{\theta} = .5\bar{x}_{1\bullet} + .5\bar{x}_{2\bullet} - \bar{x}_{3\bullet} = -.527$ and $\Sigma c_i^2 = 1.50$. With $t_{.025,27} = 2.052$ and MSE $= .0660$, the desired CI is (from (10.5))

$-.527 \pm (2.052)\sqrt{\dfrac{(.0660)(1.50)}{10}} = -.527 \pm .204 = (-.731, -.323)$.

19. MSTr = 140, error d.f. = 12, so $f = \dfrac{140}{SSE/12} = \dfrac{1680}{SSE}$ and $F_{.05,2,12} = 3.89$.

$w = Q_{.05,3,12}\sqrt{\dfrac{MSE}{J}} = 3.77\sqrt{\dfrac{SSE}{60}} = .4867\sqrt{SSE}$. Thus we wish $\dfrac{1680}{SSE} > 3.89$

(significance of f) and $.4867\sqrt{SSE} > 10$ ($= 20 - 10$, the difference between the extreme $\bar{x}_{i\bullet}$'s - so no significant differences are identified). These become $431.88 > SSE$ and $SSE > 422.16$, so SSE = 425 will work.

21.

a. Grand mean = 222.167, MSTr = 38,015.1333, MSE = 1,681.8333, and f = 22.6. The hypotheses are $H_0 : \mu_1 = ... = \mu_6$ vs. H_a :at least two μ_i's differ . Reject H_o if $f \geq F_{.01,5,78}$ (but since there is no table value for $v_2 = 78$, use $f \geq F_{.01,5,60} = 3.34$) With $22.6 \geq 3.34$, we reject H_o. The data indicates there is a dependence on injection regimen.

Chapter 11: The Analysis of Variance

b. Assume $t_{.005,78} \approx 2.645$

    i)   Confidence interval for $\mu_1 - \frac{1}{5}(\mu_2 + \mu_3 + \mu_4 + \mu_5 + \mu_6)$:

$$\Sigma c_i \bar{x}_i \pm t_{\alpha/2, J(J-1)} \sqrt{\frac{MSE(\Sigma c_i^2)}{J}}$$

$$= -67.4 \pm (2.645)\sqrt{\frac{1,681.8333(1.2)}{14}} = (-99.16, -35.64).$$

    ii)   Confidence interval for $\frac{1}{4}(\mu_2 + \mu_3 + \mu_4 + \mu_5) - \mu_6$:

$$= 61.75 \pm (2.645)\sqrt{\frac{1,681.8333(1.25)}{14}} = (29.34, 94.16)$$

23. $J_1 = 5$, $J_2 = 4$, $J_3 = 4$, $J_4 = 5$, $\bar{x}_{1\bullet} = 58.28$, $\bar{x}_{2\bullet} = 55.40$, $\bar{x}_{3\bullet} = 50.85$, $\bar{x}_{4\bullet} = 45.50$, MSE = 8.89. With

$$W_{ij} = Q_{.05,4,14} \cdot \sqrt{\frac{MSE}{2}\left(\frac{1}{J_i} + \frac{1}{J_j}\right)} = 4.11\sqrt{\frac{8.89}{2}\left(\frac{1}{J_i} + \frac{1}{J_j}\right)},$$

$\bar{x}_{1\bullet} - \bar{x}_{2\bullet} \pm W_{12} = (2.88) \pm (5.81)$; $\bar{x}_{1\bullet} - \bar{x}_{3\bullet} \pm W_{13} = (7.43) \pm (5.81)*$;

$\bar{x}_{1\bullet} - \bar{x}_{4\bullet} \pm W_{14} = (12.78) \pm (5.48)*$; $\bar{x}_{2\bullet} - \bar{x}_{3\bullet} \pm W_{23} = (4.55) \pm (6.13)$;

$\bar{x}_{2\bullet} - \bar{x}_{4\bullet} \pm W_{24} = (9.90) \pm (5.81)*$; $\bar{x}_{3\bullet} - \bar{x}_{4\bullet} \pm W_{34} = (5.35) \pm (5.81)$;

*Indicates an interval that doesn't include zero, corresponding to μ's that are judged significantly different.

This underscoring pattern does not have a very straightforward interpretation.

Chapter 11: The Analysis of Variance

25.

a. The distributions of the polyunsaturated fat percentages for each of the four regimens must be normal with equal variances.

b. We have all the $\overline{X}_{i.}$'s, and we need the grand mean:

$$\overline{X}_{..} = \frac{8(43.0)+13(42.4)+17(43.1)+14(43.5)}{52} = \frac{2236.9}{52} = 43.017$$

$$SSTr = \sum J_i(\overline{x}_{i.} - \overline{x}_{..})^2 = 8(43.0 - 43.017)^2 + 13(42.4 - 43.017)^2$$

$$+ 17(43.1 - 43.017)^2 + 13(43.5 - 43.017)^2 = 8.334$$

and $MSTr = \dfrac{8.334}{3} = 2.778$

$$SSTr = \sum (J_i - 1)s^2 = 7(1.5)^2 + 12(1.3)^2 + 16(1.2)^2 + 13(1.2)^2 = 77.79$$

and $MSE = \dfrac{77.79}{48} = 1.621$. Then $f = \dfrac{MSTr}{MSE} = \dfrac{2.778}{1.621} = 1.714$ Since

$1.714 < F_{.10,3,50} = 2.20$, we can say that the p-value is $> .10$. We do not reject the null hypothesis at significance level .10 (or any smaller), so we conclude that the data suggests no difference in the percentages for the different regimens.

27.

a. Let μ_i = true average folacin content for specimens of brand I. The hypotheses to be tested are $H_0 : \mu_1 = \mu_2 = \mu_3 = \mu_4$ vs. H_a : at least two μ_i's differ .

$$\Sigma\Sigma x_{ij}^2 = 1246.88 \text{ and } \frac{x_{..}^2}{n} = \frac{(168.4)^2}{24} = 1181.61, \text{ so SST} = 65.27.$$

$$\frac{\Sigma x_{i\bullet}^2}{J_i} = \frac{(57.9)^2}{7} + \frac{(37.5)^2}{5} + \frac{(38.1)^2}{6} + \frac{(34.9)^2}{6} = 1205.10, \text{ so}$$

$$SSTr = 1205.10 - 1181.61 = 23.49.$$

Source	Df	SS	MS	F
Treatments	3	23.49	7.83	3.75
Error	20	41.78	2.09	
Total	23	65.27		

With numerator df = 3 and denominator = 20,

$F_{.05,3,20} = 3.10 < 3.75 < F_{.01,3,20} = 4.94$, so $.01 < p - value < .05$, and since the

Chapter 11: The Analysis of Variance

p-value < .05, we reject H_o. At least one of the pairs of brands of green tea has different average folacin content.

b. With $\bar{x}_{i\bullet}$ = 8.27, 7.50, 6.35, and 5.82 for i = 1, 2, 3, 4, we calculate the residuals $x_{ij} - \bar{x}_{i\bullet}$ for all observations. A normal probability plot appears below, and indicates that the distribution of residuals could be normal, so the normality assumption is plausible. The sample standard deviations are 1.463, 1.681, 1.060, and 1.551, so the equal variance assumption is plausible (since the largest sd is less than twice the smallest sd).

Normal Probability Plot for ANOVA Residuals

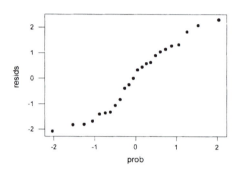

c. $Q_{.05,4,20}$ = 3.96 and $W_{ij} = 3.96 \cdot \sqrt{\dfrac{2.09}{2}\left(\dfrac{1}{J_i} + \dfrac{1}{J_j}\right)}$, so the Modified Tukey intervals are:

Pair	Interval	Pair	Interval
1,2	$.77 \pm 2.37$	2,3	1.15 ± 2.45
1,3	1.92 ± 2.25	2,4	1.68 ± 2.45
1,4	2.45 ± 2.25 *	3,4	$.53 \pm 2.34$

4	3	2	1

Only Brands 1 and 4 are significantly different from each other.

Chapter 11: The Analysis of Variance

29.
$$E(SSTr) = E\left(\sum_i J_i \bar{X}_{i\bullet}^2 - n\bar{X}_{\bullet\bullet}^2\right) = \sum_i J_i E\left(\bar{X}_{i\bullet}^2\right) - nE\left(\bar{X}_{\bullet\bullet}^2\right)$$

$$= \sum_i J_i \left[Var\left(\bar{X}_{i\bullet}\right) + \left(E\left(\bar{X}_{i\bullet}\right)\right)^2\right] - n\left[Var\left(\bar{X}_{\bullet\bullet}\right) + \left(E\left(\bar{X}_{\bullet\bullet}\right)\right)^2\right]$$

$$= \sum_i J_i \left[\frac{\sigma^2}{J_i} + \mu_i^2\right] - n\left[\frac{\sigma^2}{n} + \left(\frac{\sum_i J_i \mu_i}{n}\right)^2\right]$$

$$= I\sigma^2 + \sum_i J_i (\mu + \alpha_i)^2 - \sigma^2 - \frac{1}{n}\left[\sum_i J_i (\mu + \alpha_i)\right]^2$$

$$= (I-1)\sigma^2 + \sum_i J_i \mu^2 + 2\mu\sum_i J_i \alpha_i + \sum_i J_i \alpha_i^2 - \frac{1}{n}\left[n\mu + 0\right]^2$$

$$= (I-1)\sigma^2 + \mu^2 n + 2\mu 0 + \sum_i J_i \alpha_i^2 - n\mu^2 = (I-1)\sigma^2 + \sum_i J_i \alpha_i^2 ,$$

from which E(MSTr) is obtained through division by (I-1).

31. With $\sigma = 1$ (any other σ would yield the same Φ), $\alpha_1 = -1$, $\alpha_2 = \alpha_3 = 0$,

$\alpha_4 = 1$, $\Phi^2 = \dfrac{1\left(5(-1)^2 + 4(0)^2 + 4(0)^2 + 5(1)^2\right)}{4} = 2.5$, $\Phi = 1.58$, $v_1 = 3$,

$v_2 = 14$, and power $\approx .65$.

33. $g(x) = x\left(1 - \dfrac{x}{n}\right) = nu(1-u)$ where $u = \dfrac{x}{n}$, so $h(x) = \int [u(1-u)]^{-1/2} du$.

From a table of integrals, this gives $h(x) = \arcsin\left(\sqrt{u}\right) = \arcsin\left(\sqrt{\dfrac{x}{n}}\right)$ as the

appropriate transformation.

35.

    a. $MSA = \dfrac{30.6}{4} = 7.65$, $MSE = \dfrac{59.2}{12} = 4.93$, $f_A = \dfrac{7.65}{4.93} = 1.55$. Since

    1.55 is not $\geq F_{.05,4,12} = 3.26$, don't reject H_{oA}. There is no significant
difference in true average tire lifetime due to different makes of cars.

    b. $MSB = \dfrac{44.1}{3} = 14.70$, $f_B = \dfrac{14.70}{4.93} = 2.98$. Since 2.98 is not

    $\geq F_{.05,3,12} = 3.49$, don't reject H_{oB}. There is no significant difference in true
average tire lifetime due to different brands of tires.

Chapter 11: The Analysis of Variance

37. Direct calculation gives $\sum\sum x_{ij}^2 = 44614.21$, SST = 15653.56, SSA = 1428.28, SSB = 13444.63, and SSE = 780.65. The complete ANOVA table is below.

Source	Df	SS	MS	F
A (stimuli)	5	1428.28	285.66	5.49
B (subjects)	3	13444.63	4481.54	86.12
Error	15	780.65	52.04	
Total	23	15653.56		

Since $F_{.05,5,15} = 2.90$ and $5.49 \geq 2.90$, we conclude there are differences in the true average responses associated with different stimuli. For Tukey's procedure, $w = 4.59\sqrt{52.04/4} = 16.56$. The resulting line summary, below, shows both L1 and L2 are significantly different from L2+T, and there are no other significant differences among the stimuli. Differences among the subject means are not very important here. The normal plot of residuals shows no reason to doubt normality. However, the plot of residuals against the fitted values shows some dependence of the variance on the mean. If logged response is used in place of response, the plots look good and the F test result is similar but stronger. Furthermore, the logged response gives more significant differences in the multiple comparisons analysis.

L1	L2	T	L1+L2	L1+T	L2+T
24.8	27.9	29.1	40.3	41.2	45.1

39.

Source	Df	SS	MS	F
Angle	3	58.16	19.3867	2.5565
Connector	4	246.97	61.7425	8.1419
Error	12	91.00	7.5833	
Total	19	396.13		

$H_0 : \alpha_1 = \alpha_2 = \alpha_3 = \alpha_4 = 0$; $\qquad H_a$: at least one α_i is not zero.

$f_A = 2.5565 < F_{.01,3,12} = 5.95$, so fail to reject H_0. The data fails to indicate any effect due to the angle of pull.

Chapter 11: The Analysis of Variance

41.

a. CF = 140,454, SST = 3476, $SSTr = \dfrac{(905)^2 + (913)^2 + (936)^2}{18} - 140,454 = 28.78$,

$SSBl = \dfrac{430,295}{3} - 140,454 = 2977.67$, SSE = 469.55, MSTr = 14.39, MSE =

13.81, $f_{Tr} = 1.04$, which is clearly insignificant when compared to $F_{.05,2,34}$.

b. $f_{Bl} = 12.68$, which is significant, and suggests substantial variation among subjects. If we had not controlled for such variation, it might have affected the analysis and conclusions.

43.

Source	Df	SS	MS	f
Method	2	23.23	11.61	8.69
Batch	9	86.79	9.64	7.22
Error	18	24.04	1.34	
Total	29	134.07		

$F_{.01,2,18} = 6.01 < 8.69 < F_{.001,2,18} = 10.39$, so .001 < p-value < .01, which is significant. At least two of the curing methods produce differing average compressive strengths. (With p-value < .001, there are differences between batches as well.)

$Q_{.05,3,18} = 3.61$; $w = (3.61)\sqrt{\dfrac{1.34}{10}} = 1.32$

Method A	Method B	Method C
29.49	31.31	31.40

Methods B and C produce strengths that are not significantly different, but Method A produces strengths that are different (less) than those of both B and C.

45. $MSB = \dfrac{113.5}{4} = 28.38$, $MSE = \dfrac{25.6}{8} = 3.20$, $f_B = 8.87$, $F_{.01,4,8} = 7.01$,

and since $8.87 \geq 7.01$, we reject H_0 and conclude that $\sigma_B^2 > 0$.

Chapter 11: The Analysis of Variance

47.

$$E\left(\overline{X}_{i\bullet} - \overline{X}_{\bullet\bullet}\right) = E\left(\overline{X}_{i\bullet}\right) - E\left(\overline{X}_{\bullet\bullet}\right) = \frac{1}{J}E\left(\sum_j X_{ij}\right) - \frac{1}{IJ}E\left(\sum_i \sum_j X_{ij}\right)$$

$$= \frac{1}{J}\sum_j \left(\mu + \alpha_i + \beta_j\right) - \frac{1}{IJ}\sum_i \sum_j \left(\mu + \alpha_i + \beta_j\right)$$

$$= \mu + \alpha_i + \frac{1}{J}\sum_j \beta_j - \mu - \frac{1}{I}\sum_i \alpha_i - \frac{1}{J}\sum_j \beta_j = \alpha_i \text{ , as desired.}$$

49.

a.

Source	Df	SS	MS	f
A	2	30,763.0	15,381.50	3.79
B	3	34,185.6	11,395.20	2.81
AB	6	43,581.2	7263.53	1.79
Error	24	97,436.8	4059.87	
Total	35	205,966.6		

b. $f_{AB} = 1.79$ which is not $\geq F_{.05,6,24} = 2.51$, so H_{oAB} cannot be rejected, and we conclude that no interaction is present.

c. $f_A = 3.79$ which is $\geq F_{.05,2,24} = 3.40$, so H_{oA} is rejected at level .05.

d. $f_B = 2.81$ which is not $\geq F_{.05,3,24} = 3.01$, so H_{oB} is not rejected.

e. $Q_{.05,3,24} = 3.53$, $w = 3.53\sqrt{\dfrac{4059.87}{12}} = 64.93$.

	3	1	2
	3960.02	4010.88	4029.10

Only times 2 and 3 yield significantly different strengths.

Chapter 11: The Analysis of Variance

51.

Source	Df	SS	MS	f	$F_{.05}$	$F_{.01}$
Formulation	1	2,253.44	2,253.44	376.2**	4.75	9.33
Speed	2	230.81	115.41	19.27**	3.89	6.93
Formulation & Speed	2	18.58	9.29	1.55	3.89	6.93
Error	12	71.87	5.99			
Total	17	2,574.7				

a. There appears to be no interaction between the two factors.

b. Both formulation and speed appear to have a highly statistically significant effect on yield.

c. Let formulation = Factor A and speed = Factor B.

For Factor A: $\mu_{1\bullet} = 187.03$ $\mu_{2\bullet} = 164.66$

For Factor B: $\mu_{\bullet 1} = 177.83$ $\mu_{\bullet 2} = 170.82$ $\mu_{\bullet 3} = 178.88$

For Interaction: $\mu_{11} = 189.47$ $\mu_{12} = 180.6$ $\mu_{13} = 191.03$

$\mu_{21} = 166.2$ $\mu_{22} = 161.03$ $\mu_{33} = 166.73$

overall mean: $\mu = 175.84$

$\alpha_i = \mu_{i\bullet} - \mu:$ $\alpha_1 = 11.19$ $\alpha_2 = -11.18$

$\beta_j = \mu_{\bullet j} - \mu:$ $\beta_1 = 1.99$ $\beta_2 = -5.02$ $\beta_3 = 3.04$

$y_{ij} = \mu_{ij} - (\mu + \alpha_i + \beta_j):$

$y_{11} = .45$ $y_{12} = -1.41$ $y_{13} = .96$

$y_{21} = -.45$ $y_{22} = 1.39$ $y_{23} = -.97$

d.

Observed	Fitted	Residual	Observed	Fitted	Residual
189.7	189.47	0.23	161.7	161.03	0.67
188.6	189.47	-0.87	159.8	161.03	-1.23
190.1	189.47	0.63	161.6	161.03	0.57
165.1	166.2	-1.1	189.0	191.03	-2.03
165.9	166.2	-0.3	193.0	191.03	1.97
167.6	166.2	1.4	191.1	191.03	0.07
185.1	180.6	4.5	163.3	166.73	-3.43
179.4	180.6	-1.2	166.6	166.73	-0.13
177.3	180.6	-3.3	170.3	166.73	3.57

e.

i	Residual	Percentile	z-percentile
1	-3.43	2.778	-1.91
2	-3.30	8.333	-1.38
3	-2.03	13.889	-1.09
4	-1.23	19.444	-0.86
5	-1.20	25.000	-0.67
6	-1.10	30.556	-0.51
7	-0.87	36.111	-0.36
8	-0.30	41.667	-0.21
9	-0.13	47.222	-0.07
10	0.07	52.778	0.07
11	0.23	58.333	0.21
12	0.57	63.889	0.36
13	0.63	69.444	0.51
14	0.67	75.000	0.67
15	1.40	80.556	0.86
16	1.97	86.111	1.09
17	3.57	91.667	1.38
18	4.50	97.222	1.91

Normal Probability Plot of ANOVA Residuals

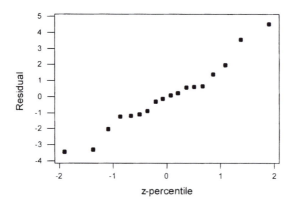

The residuals appear to be normally distributed.

Chapter 11: The Analysis of Variance

53. The relevant null hypotheses are $H_{0A} : \alpha_1 = \alpha_2 = \alpha_3 = \alpha_4 = 0$; $H_{0B} : \sigma_B^2 = 0$; $H_{0AB} : \sigma_G^2 = 0$.

$$SST = 11,499,492 - \frac{(16,598)^2}{24} = 20,591.83 \, ,$$

$$SSE = 11,499,492 - \frac{(22,982,552)}{2} = 8216.0 \, ,$$

$$SSA = \left[\frac{(4112)^2 + (4227)^2 + (4122)^2 + (4137)^2}{6} \right] - \frac{(16,598)^2}{24} = 1387.5 \, ,$$

$$SSB = \left[\frac{(5413)^2 + (5621)^2 + (5564)^2}{8} \right] - \frac{(16,598)^2}{24} = 2888.08 \, ,$$

$$SSAB = 20,591.83 - [8216.0 + 1387.5 + 2888.08] = 8216.25$$

Source	Df	SS	MS	f	$F_{.05}$
A	3	1,387.5	462.5	$\frac{MSA}{MSAB} = .34$	4.76
B	2	2,888.08	1,444.04	$\frac{MSB}{MSAB} = 1.07$	5.14
AB	6	8,100.25	1,350.04	$\frac{MSAB}{MSE} = 1.97$	3.00
Error	12	8,216.0	684.67		
Total	23	20,591.83			

Interaction between brand and writing surface has no significant effect on the lifetime of the pen, and since neither f_A nor f_B is greater than its respective critical value, we can conclude that neither the surface nor the brand of pen has a significant effect on the writing lifetime.

55.

a. $E\left(\overline{X}_{i..} - \overline{X}_{...}\right) = \frac{1}{JK} \sum_j \sum_k E\left(X_{ijk}\right) - \frac{1}{IJK} \sum_i \sum_j \sum_k E\left(X_{ijk}\right)$

$= \frac{1}{JK} \sum_j \sum_k \left(\mu + \alpha_i + \beta_j + \gamma_{ij}\right) - \frac{1}{IJK} \sum_i \sum_j \sum_k \left(\mu + \alpha_i + \beta_j + \gamma_{ij}\right) = \mu + \alpha_i - \mu = \alpha_i$

b. $E\left(\hat{\gamma}_{ij}\right) = \frac{1}{K} \sum_k E\left(X_{ijk}\right) - \frac{1}{JK} \sum_j \sum_k E\left(X_{ijk}\right) - \frac{1}{IK} \sum_i \sum_k E\left(X_{ijk}\right) + \frac{1}{IJK} \sum_i \sum_j \sum_k E\left(X_{ijk}\right)$

$= \mu + \alpha_i + \beta_j + \gamma_{ij} - \left(\mu + \alpha_i\right) - \left(\mu + \beta_j\right) + \mu = \gamma_{ij}$

Chapter 11: The Analysis of Variance

57.

a. $\dfrac{E(MSAB)}{E(MSE)} = 1 + \dfrac{K\sigma_G^2}{\sigma^2} = 1$ if $\sigma_G^2 = 0$ and > 1 if $\sigma_G^2 > 0$, so $\dfrac{MSAB}{MSE}$ is the appropriate F ratio.

b. $\dfrac{E(MSA)}{E(MSAB)} = \dfrac{\sigma^2 + K\sigma_G^2 + JK\sigma_A^2}{\sigma^2 + K\sigma_G^2} = 1 + \dfrac{JK\sigma_A^2}{\sigma^2 + K\sigma_G^2} = 1$ if $\sigma_A^2 = 0$ and > 1 if $\sigma_A^2 > 0$, so $\dfrac{MSA}{MSAB}$ is the appropriate F ratio.

59.

a. $H_0 : \mu_1 = \ldots = \mu_5$ will be rejected in favor of H_a : at least two μ_i's differ if $f \geq F_{.05,4,40} = 2.61$. With $\bar{x}_{\bullet\bullet} = 30.82$, straightforward calculation yields

$MSTr = \dfrac{221.112}{4} = 55.278$, $MSE = \dfrac{80.4591}{5} = 16.1098$, and

$f = \dfrac{55.278}{16.1098} = 3.43$. Because $3.43 \geq 2.61$, H_0 is rejected. There is a difference among the five teaching methods with respect to true mean exam score.

b. The format of this test is identical to that of part **a**. The calculated test statistic is $f = \dfrac{33.12}{20.109} = 1.65$. Since $1.65 < 2.61$, H_0 is not rejected. The data suggests that with respect to true average retention scores, the five methods are not different from one another.

61.

a. $x_{1\bullet} = 15.48$, $x_{2\bullet} = 15.78$, $x_{3\bullet} = 12.78$, $x_{4\bullet} = 14.46$, $x_{5\bullet} = 14.94$
$x_{\bullet\bullet} = 73.44$, so $CF = 179.78$, SST = 3.62, SSTr = $180.71 - 179.78 = .93$,
SSE = 3.62 - .93 = 2.69.

Source	Df	SS	MS	F
Treatments	4	.93	.233	2.16
Error	25	2.69	.108	
Total	29	3.62		

$F_{.05,4,25} = 2.76$. Since 2.16 is not ≥ 2.76, do not reject H_0 at level .05.

Chapter 11: The Analysis of Variance

b. $\hat{\theta} = 2.58 - \dfrac{2.63 + 2.13 + 2.41 + 2.49}{4} = .165$, $t_{.025,25} = 2.060$, MSE = .108, and

$\Sigma c_i^2 = (1)^2 + (-.25)^2 + (-.25)^2 + (-.25)^2 + (-.25)^2 = 1.25$, so a 95%
confidence interval for θ is

$.165 \pm 2.060\sqrt{\dfrac{(.108)(1.25)}{6}} = .165 \pm .309 = (-.144, .474)$. This interval does

include zero, so 0 is a plausible value for θ.

c. $\mu_1 = \mu_2 = \mu_3, \mu_4 = \mu_5 = \mu_1 - \sigma$, so $\mu = \mu_1 - \frac{2}{5}\sigma$,

$\alpha_1 = \alpha_2 = \alpha_3 = \frac{2}{5}\sigma$, $\alpha_4 = \alpha_5 = -\frac{3}{5}\sigma$. Then $\Phi^2 = \dfrac{J}{I}\sum\dfrac{\alpha_i^2}{\sigma^2}$

$= \dfrac{6}{5}\left[\dfrac{3\left(\frac{2}{5}\sigma\right)^2}{\sigma^2} + \dfrac{2\left(-\frac{3}{5}\sigma\right)^2}{\sigma^2}\right] = 1.632$ and $\Phi = 1.28$, $v_1 = 4$, $v_2 = 25$.

By inspection of figure (10.6), power $\approx .48$, so $\beta \approx .52$

63.

a. μ_i = true average CFF for the three iris colors. Then the hypotheses are
$H_0 : \mu_1 = \mu_2 = \mu_3$ vs. H_a : at least two $\mu_i's$ differ. SST = 13,659.67 –
13,598.36 = 61.31,

$SSTR = \left(\dfrac{(204.7)^2}{8} + \dfrac{(134.6)^2}{5} + \dfrac{(169.0)^2}{6}\right) - 13,598.36 = 23.00$ The ANOVA

table follows:

Source	Df	SS	MS	F
Treatments	2	23.00	11.50	4.803
Error	16	38.31	2.39	
Total	18	61.31		

Because $F_{.05,2,16} = 3.63 < 4.803 < F_{.01,2,16} = 6.23$, .01 < p-value < .05, so
we reject H_0. There are differences in CFF based on iris color.

b. $Q_{.05,3,16} = 3.65$ and $W_{ij} = 3.65 \cdot \sqrt{\dfrac{2.39}{2}\left(\dfrac{1}{J_i} + \dfrac{1}{J_j}\right)}$, so the Modified Tukey

intervals are:

Pair	$\left(\bar{x}_{i\bullet} - \bar{x}_{j\bullet}\right) \pm W_{ij}$
1,2	-1.33 ± 2.27
1,3	-2.58 ± 2.15 *
2,3	-1.25 ± 2.42

Brown	Green	Blue
25.59	26.92	28.17

The CFF is only significantly different for Brown and Blue iris color.

65.

Source	Df	SS	MS	F	$F_{.05}$
Treatments	3	24,937.63	8312.54	1117.8	4.07
Error	8	59.49	7.44		
Total	11	24,997.12			

Because $1117.8 \geq 4.07$, $H_0 : \mu_1 = \mu_2 = \mu_3 = \mu_4$ is rejected.

$Q_{.05,4,8} = 4.53$, so $w = 4.53\sqrt{\dfrac{7.44}{3}} = 7.13$. The four sample means are

$\bar{x}_{4\bullet} = 29.92$, $\bar{x}_{1\bullet} = 33.96$, $\bar{x}_{3\bullet} = 115.84$, and $\bar{x}_{2\bullet} = 129.30$. Only

$\bar{x}_{1\bullet} - \bar{x}_{4\bullet} < 7.13$, so all means are judged significantly different from one another

except for μ_4 and μ_1 (corresponding to PCM and OCM).

67. The ordered residuals are –6.67, -5.67, -4, -2.67, -1, -1, 0, 0, 0, .33, .33, .33, 1, 1, 2.33, 4, 5.33, 6.33. The corresponding z percentiles are –1.91, -1.38, -1.09, -.86, -.67, -.51, -.36, -.21, -.07, .07, .21, .36, .51, .67, .86, 1.09, 1.38, and 1.91. The resulting plot of the respective pairs (the Normal Probability Plot) is reasonably straight, and thus there is no reason to doubt the normality assumption.

69.

Source	Df	SS	MS	f
A	1	322.667	322.667	980.38
B	3	35.623	11.874	36.08
AB	3	8.557	2.852	8.67
Error	16	5.266	.329	
Total	23	372.113		

We first test the null hypothesis of no interactions ($H_0 : \gamma_{ij} = 0$ for all I, j). H$_o$ will be rejected at level .05 if $f_{AB} = \dfrac{MSAB}{MSE} \geq F_{.05,3,16} = 3.24$. Because $8.67 \geq 3.24$, H$_o$ is rejected. Because we have concluded that interaction is present, tests for main effects are not appropriate.

Chapter 12: Regression and Correlation

1.

    a. Stem and Leaf display of temp:

```
17|0
17|23              stem = tens
17|445             leaf = ones
17|67
17|
18|0000011
18|2222
18|445
18|6
18|8
```

180 appears to be a typical value for this data. The distribution is reasonably symmetric in appearance and somewhat bell-shaped. The variation in the data is fairly small since the range of values (188 – 170 = 18) is fairly small compared to the typical value of 180.

```
0|889
1|0000             stem = ones
1|3                leaf = tenths
1|4444
1|66
1|8889
2|11
2|
2|5
2|6
2|
3|00
```

For the ratio data, a typical value is around 1.6 and the distribution appears to be positively skewed. The variation in the data is large since the range of the data (3.08 - .84 = 2.24) is very large compared to the typical value of 1.6. The two largest values could be outliers.

    b. The efficiency ratio is not uniquely determined by temperature since there are several instances in the data of equal temperatures associated with different efficiency ratios. For example, the five observations with temperatures of 180 each have different efficiency ratios.

Chapter 12: Regression and Correlation

c. A scatter plot of the data appears below. The points exhibit quite a bit of variation and do not appear to fall close to any straight line or simple curve.

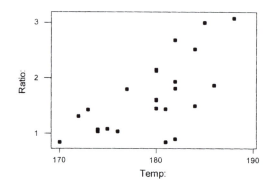

3. A scatter plot of the data appears below. The points fall very close to a straight line with an intercept of approximately 0 and a slope of about 1. This suggests that the two methods are producing substantially the same concentration measurements.

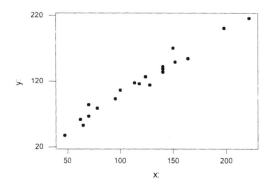

Chapter 12: Regression and Correlation

5.

    a.   The scatter plot with axes intersecting at (0,0) is shown below.

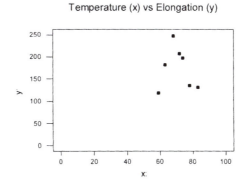

Temperature (x) vs Elongation (y)

    b.   The scatter plot with axes intersecting at (55, 100) is shown below.

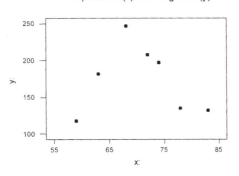

Temperature (x) vs Elongation (y)

    c.   A parabola appears to provide a good fit to both graphs.

Chapter 12: Regression and Correlation

7.

    a. $\mu_{Y \cdot 2500} = 1800 + 1.3(2500) = 5050$

    b.   expected change = slope = $\beta_1 = 1.3$

    c.   expected change = $100\beta_1 = 130$

    d.   expected change = $-100\beta_1 = -130$

9.

    a. β_1 = expected change in flow rate (y) associated with a one inch increase in pressure drop (x) = .095.

    b.   We expect flow rate to decrease by $5\beta_1 = .475$.

    c. $\mu_{Y \cdot 10} = -.12 + .095(10) = .83$, and $\mu_{Y \cdot 15} = -.12 + .095(15) = 1.305$.

    d. $P(Y > .835) = P\left(Z > \dfrac{.835 - .830}{.025}\right) = P(Z > .20) = .4207$

$$P(Y > .840) = P\left(Z > \frac{.840 - .830}{.025}\right) = P(Z > .40) = .3446$$

    e.   Let Y_1 and Y_2 denote pressure drops for flow rates of 10 and 11, respectively. Then $\mu_{Y \cdot 11} = .925$, so $Y_1 - Y_2$ has expected value .830 - .925 = -.095, and s.d.

$$\sqrt{(.025)^2 + (.025)^2} = .035355.\ \text{Thus}$$

$$P(Y_1 > Y_2) = P(Y_1 - Y_2 > 0) = P\left(z > \frac{+.095}{.035355}\right) = P(Z > 2.69) = .0036$$

11.

    a. β_1 = expected change for a one degree increase = -.01, and $10\beta_1 = -.1$ is the expected change for a 10 degree increase.

    b. $\mu_{Y \cdot 200} = 5.00 - .01(200) = 3$, and $\mu_{Y \cdot 250} = 2.5$.

    c.   The probability that the first observation is between 2.4 and 2.6 is

$$P(2.4 \le Y \le 2.6) = P\left(\frac{2.4 - 2.5}{.075} \le Z \le \frac{2.6 - 2.5}{.075}\right) = P(-1.33 \le Z \le 1.33) = .8164.$$

The probability that any particular one of the other four observations is between 2.4 and 2.6 is also .8164, so the probability that all five are between 2.4 and 2.6 is $(.8164)^5 = .3627$.

Chapter 12: Regression and Correlation

d. Let Y_1 and Y_2 denote the times at the higher and lower temperatures, respectively. Then $Y_1 - Y_2$ has expected value $5.00 - .01(x+1) - (5.00 - .01x) = -.01$. The standard deviation of $Y_1 - Y_2$ is

$$\sqrt{(.075)^2 + (.075)^2} = .10607 . \text{ Thus}$$

$$P(Y_1 - Y_2 > 0) = P\left(z > \frac{-(-.01)}{.10607}\right) = P(Z > .09) = .4641 .$$

13.

a. $S_{xx} = 39,095 - \dfrac{(517)^2}{14} = 20,002.929$, $S_{xy} = 25,825 - \dfrac{(517)(346)}{14} = 13047.714$;

$\hat{\beta}_1 = \dfrac{S_{xy}}{S_{xx}} = \dfrac{13,047.714}{20,002.929} = .652$; $\hat{\beta}_0 = \dfrac{\Sigma y - \hat{\beta}_1 \Sigma x}{n} = \dfrac{346 - (.652)(517)}{14} = .626$, so

the equation of the least squares regression line is $y = .626 + .652x$.

b. $\hat{y}_{(35)} = .626 + .652(35) = 23.456$. The residual is $y - \hat{y} = 21 - 23.456 = -2.456$.

c. $S_{yy} = 17,454 - \dfrac{(346)^2}{14} = 8902.857$, so

$SSE = 8902.857 - (.652)(13047.714) = 395.747$.

$$\hat{\sigma} = \sqrt{\dfrac{SSE}{n-2}} = \sqrt{\dfrac{395.747}{12}} = 5.743 .$$

d. $SST = S_{yy} = 8902.857$; $r^2 = 1 - \dfrac{SSE}{SST} = 1 - \dfrac{395.747}{8902.857} = .956$.

e. Without the two upper extreme observations, the new summary values are $n = 12, \Sigma x = 272, \Sigma x^2 = 8322, \Sigma y = 181, \Sigma y^2 = 3729, \Sigma xy = 5320$. The new $S_{xx} = 2156.667, S_{yy} = 998.917, S_{xy} = 1217.333$. New $\hat{\beta}_1 = .56445$ and $\hat{\beta}_0 = 2.2891$, which yields the new equation $y = 2.2891 + .56445x$. Removing the two values changes the position of the line considerably, and the slope slightly. The new $r^2 = 1 - \dfrac{311.79}{998.917} = .6879$, which is much worse than that of the original set of observations.

Chapter 12: Regression and Correlation

15.

a. $n = 24$, $\Sigma x_i = 4308$, $\Sigma y_i = 40.09$, $\Sigma x_i^2 = 773{,}790$, $\Sigma y_i^2 = 76.8823$,

$\Sigma x_i y_i = 7{,}243.65$. $S_{xx} = 773{,}790 - \dfrac{(4308)^2}{24} = 504.0$,

$S_{yy} = 76.8823 - \dfrac{(40.09)^2}{24} = 9.9153$, and

$S_{xy} = 7{,}243.65 - \dfrac{(4308)(40.09)}{24} = 45.8246$. $\hat{\beta}_1 = \dfrac{S_{xy}}{S_{xx}} = \dfrac{45.8246}{504} = .09092$

and $\hat{\beta}_0 = \dfrac{40.09}{24} - (.09092)\dfrac{4308}{24} = -14.6497$. The equation of the estimated

regression line is $\hat{y} = -14.6497 + .09092x$.

b. When x = 182, $\hat{y} = -14.6497 + .09092(182) = 1.8997$. So when the tank
temperature is 182, we would predict an efficiency ratio of 1.8997.

c. The four observations for which temperature is 182 are: (182, .90), (182, 1.81),
(182, 1.94), and (182, 2.68). Their corresponding residuals are:
$.90 - 1.8997 = -0.9977$, $1.81 - 1.8997 = -0.0877$,
$1.94 - 1.8997 = 0.0423$, $2.68 - 1.8997 = 0.7823$. These residuals do not
all have the same sign because in the cases of the first two pairs of observations,
the observed efficiency ratios were smaller than the predicted value of 1.8997.
Whereas, in the cases of the last two pairs of observations, the observed
efficiency ratios were larger than the predicted value.

d. $SSE = S_{yy} - \hat{\beta}_1 S_{xy} = 9.9153 - (.09092)(45.8246) = 5.7489$.

$r^2 = 1 - \dfrac{SSE}{SST} = 1 - \dfrac{5.7489}{9.9153} = .4202$. (42.02% of the observed variation in

efficiency ratio can be attributed to the approximate linear relationship between
the efficiency ratio and the tank temperature.)

17.

   a.

Rainfall volume (x) vs Runoff volume (y)

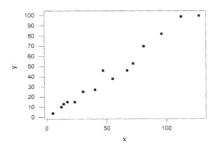

Yes, the scatterplot shows a strong linear relationship between rainfall volume and runoff volume, thus it supports the use of the simple linear regression model.

   b. $\bar{x} = 53.200$, $\bar{y} = 42.867$, $S_{xx} = 63040 - \dfrac{(798)^2}{15} = 20{,}586.4$,

$S_{yy} = 41{,}999 - \dfrac{(643)^2}{15} = 14{,}435.7$, and

$S_{xy} = 51{,}232 - \dfrac{(798)(643)}{15} = 17{,}024.4$. $\hat{\beta}_1 = \dfrac{S_{xy}}{S_{xx}} = \dfrac{17{,}024.4}{20{,}586.4} = .82697$

and $\hat{\beta}_0 = 42.867 - (.82697)53.2 = -1.1278$.

   c. $\mu_{y\cdot 50} = -1.1278 + .82697(50) = 40.2207$.

   d. $SSE = S_{yy} - \hat{\beta}_1 S_{xy} = 14{,}435.7 - (.82697)(17{,}324.4) = 357.07$.

$s = \hat{\sigma} = \sqrt{\dfrac{SSE}{n-2}} = \sqrt{\dfrac{357.07}{13}} = 5.24$.

   e. $r^2 = 1 - \dfrac{SSE}{SST} = 1 - \dfrac{357.07}{14{,}435.7} = .9753$. So 97.53% of the observed variation in

runoff volume can be attributed to the simple linear regression relationship between runoff and rainfall.

Chapter 12: Regression and Correlation

19. $n = 14$, $\Sigma x_i = 3300$, $\Sigma y_i = 5010$, $\Sigma x_i^2 = 913,750$, $\Sigma y_i^2 = 2,207,100$,

$\Sigma x_i y_i = 1,413,500$

a. $\hat{\beta}_1 = \dfrac{3,256,000}{1,902,500} = 1.71143233$, $\hat{\beta}_0 = -45.55190543$, so we use the equation

$y = -45.5519 + 1.7114x$.

b. $\hat{\mu}_{Y.225} = -45.5519 + 1.7114(225) = 339.51$

c. Estimated expected change $= -50\hat{\beta}_1 = -85.57$

d. No, the value 500 is outside the range of x values for which observations were
available (the danger of extrapolation).

e. Yes: a scatter plot shows a very strong linear relationship, and we can easily
calculate from the summary statistics that $r^2 = 96.1\%$. So, 96.1% of the variation
in emission rate can be explained by this linear model.

21.

a. The summary statistics can easily be verified using Minitab or Excel, etc.

b. $\hat{\beta}_1 = \dfrac{491.4}{744.16} = .66034186$, $\hat{\beta}_0 = -2.18247148$

c. predicted $y = \hat{\beta}_0 + \hat{\beta}_1(15) = 7.72$

d. $\hat{\mu}_{Y.15} = \hat{\beta}_0 + \hat{\beta}_1(15) = 7.72$

Chapter 12: Regression and Correlation

23. To ease the analysis, consider the log-likelihood (i.e., the natural logarithm of the joint pdf). In terms of the β's, the log-likelihood is $C - \dfrac{1}{2\sigma^2}\sum(y_i - [\beta_0 + \beta_1 x_i])^2$.

We find the MLEs by differentiating this with respect to the β's and setting these partial derivatives equal to zero. But this is identical to the derivation of the normal equations, from which we derived b_0 and b_1. Therefore, the MLEs of the β's are indeed the least squares estimates.

25. The new slope (based on y in °F) will equal $1.8b_1$, while the new intercept will equal $1.8b_0+32$. To see this, use rescaling properties: the new average (in °F) for the responses will be $1.8\bar{y}+32$, and the new covariance term will be $\sum(1.8\,y_i +32 - [1.8\bar{y}+32])(x_i-\bar{x}) = 1.8S_{xy}$. The x-values remain unchanged, and so the new slope is $1.8S_{xy}/S_{xx} = 1.8b_1$, and the new intercept is (new "$\bar{y}$") – (new slope)$\bar{x} = 1.8\bar{y}+32 - 1.8b_1\bar{x} = 1.8(\bar{y} - b_1\bar{x}) + 32 = 1.8b_0+32$.

27. We show that when $\bar{x}$ is substituted for x in $\hat{\beta}_0 + \hat{\beta}_1 x$, $\bar{y}$ results, so that $(\bar{x},\bar{y})$ is on the line $y = \hat{\beta}_0 + \hat{\beta}_1 x$: $\hat{\beta}_0 + \hat{\beta}_1\bar{x} = \dfrac{\sum y_i - \hat{\beta}_1\sum x_i}{n} + \hat{\beta}_1\bar{x} = \bar{y} - \hat{\beta}_1\bar{x} + \hat{\beta}_1\bar{x} = \bar{y}$.

29.

a. Subtracting $\bar{x}$ from each x_i shifts the plot in a rigid fashion $\bar{x}$ units to the left without otherwise altering its character. The last squares line for the new plot will thus have the same slope as the one for the old plot. Since the new line is $\bar{x}$ units to the left of the old one, the new y intercept (height at x = 0) is the height of the old line at x = $\bar{x}$, which is $\hat{\beta}_0 + \hat{\beta}_1\bar{x} = \bar{y}$ (since from exercise 20, $(\bar{x},\bar{y})$ is on the old line). Thus the new y intercept is $\bar{y}$.

b. We wish b_0 and b_1 to minimize $f(b_0, b_1) = \sum[y_i - (b_0 + b_1(x_i - \bar{x}))]^2$. Equating $\dfrac{\partial f}{\partial b_0}$ to $\dfrac{\partial f}{\partial b_1}$ to 0 yields $nb_0 + b_1\sum(x_i - \bar{x}) = \sum y_i$,

$b_0\sum(x_i - \bar{x}) + b_1\sum(x_i - \bar{x})^2 = \sum(x_i - \bar{x})^2 = \sum(x_i - \bar{x})y_i$. Since $\sum(x_i - \bar{x}) = 0$, $b_0 = \bar{y}$, and since $\sum(x_i - \bar{x})y_i = \sum(x_i - \bar{x})(y_i - \bar{y})$ [because $\sum(x_i - \bar{x})\bar{y} = \bar{y}\sum(x_i - \bar{x})$], $b_1 = \hat{\beta}_1$. Thus $\hat{\beta}_0^* = \bar{Y}$ and $\hat{\beta}_1^* = \hat{\beta}_1$.

Chapter 12: Regression and Correlation

31.

a. $\Sigma(x_i - \bar{x})^2 = 70,000$, so $V(\hat{\beta}_1) = \dfrac{.5^2}{70,000} = .0000036$ and the standard deviation

of $\hat{\beta}_1$ is $\sqrt{.00000036} = .00189$.

b. $P(.006 \le \hat{\beta}_1 \le .010) = P\left(\dfrac{.006 - .008}{.00189} \le Z \le \dfrac{.010 - .008}{.00189}\right) = P(-10.58 \le Z \le 10.58)$

$= 1$.

c. Although n = 11 here and n = 7 in **a**, $\Sigma(x_i - \bar{x})^2 = 24,750$ now, which is smaller than in **a**. Because this appears in the denominator of $V(\hat{\beta}_1)$, the variance is smaller for the choice of x values in **a**.

33.

a. Error d.f. = n – 2 = 9, $t_{.025,9} = 2.262$, and so the desired confidence interval is

$\hat{\beta}_1 \pm t_{.025,9} \cdot s_{\hat{\beta}_1} = 0.9555 \pm (2.262)(0.1971) = (.5097, 1.4013)$. We are 95%

confident that the true average difference in daughter's height associated with a 1 inch difference in midparent height is between .5097 inches and 1.4013 inches.

b. We wish to test $H_o : \beta_1 \ge 1$ vs. $H_a : \beta_1 < 1$. The calculated t statistic is

$t = \dfrac{\hat{\beta}_1 - 1}{s_{\hat{\beta}_1}} = \dfrac{0.9555 - 1}{.1971} = -0.226$, which yields a p-value of .413 at 9 d.f. Thus,

we fail to reject H_o; i.e., there is not enough evidence to contradict the prior belief.

35.

a. We want a 95% CI for β_1: $\hat{\beta}_1 \pm t_{.025,15} \cdot s_{\hat{\beta}_1}$. First, we need our point estimate,

$\hat{\beta}_1$. Using the given summary statistics, $S_{xx} = 3056.69 - \dfrac{(222.1)^2}{17} = 155.019$,

$S_{xy} = 2759.6 - \dfrac{(222.1)(193)}{17} = 238.112$, and $\hat{\beta}_1 = \dfrac{S_{xy}}{S_{xx}} = \dfrac{238.112}{115.019} = 1.536$. We

need $\hat{\beta}_0 = \dfrac{193 - (1.536)(222.1)}{17} = -8.715$ to calculate the SSE:

$SSE = 2975 - (-8.715)(193) - (1.536)(2759.6) = 418.2494$. Then

Chapter 12: Regression and Correlation

$s = \sqrt{\dfrac{418.2494}{15}} = 5.28$ and $s_{\hat{\beta}_1} = \dfrac{5.28}{\sqrt{155.019}} = .424$. With $t_{.025,15} = 2.131$, our

CI is $1.536 \pm 2.131 \cdot (.424) = (.632, 2.440)$. With 95% confidence, we estimate that the change in reported nausea percentage for every one-unit change in motion sickness dose is between .632 and 2.440.

b. We test the hypotheses $H_o : \beta_1 = 0$ vs $H_a : \beta_1 \neq 0$, and the test statistic is

$t = \dfrac{1.536}{.424} = 3.6226$. With df=15, the two-tailed p-value = 2P(t > 3.6226) = 2(

.001) = .002. With a p-value of .002, we would reject the null hypothesis at most reasonable significance levels. This suggests that there is a useful linear relationship between motion sickness dose and reported nausea.

c. No. A regression model is only useful for estimating values of nausea % when using dosages between 6.0 and 17.6 – the range of values sampled.

d. Removing the point (6.0, 2.50), the new summary stats are: $n = 16$, $\Sigma x_i = 216.1$, $\Sigma y_i = 190.5$, $\Sigma x_i^2 = 3020.69$, $\Sigma y_i^2 = 2968.75$, $\Sigma x_i y_i = 2744.6$, and then $\hat{\beta}_1 = 1.683$, $\hat{\beta}_0 = -10.826$, SSE = 411.6888, $s = 5.423$, $s_{\hat{\beta}_1} = .5371$, and the new CI is $1.683 \pm 2.145(.5371)$, or $(.531, 2.835)$. The interval is a little wider. But removing the one observation did not change it that much. The observation does not seem to exert undue influence.

37.

a. From Exercise 19, SSE = 16.205.45, so $s^2 = 1350.454$, $s = 36.75$, and

$s_{\hat{\beta}_1} = \dfrac{36.75}{368.636} = .0997$. Thus $t = \dfrac{1.711}{.0997} = 17.2 > 4.318 = t_{.0005,14}$, so p-value <

.001. Because the p-value < .01, $H_o : \beta_1 = 0$ is rejected at level .01 in favor of the conclusion that the model is useful $(\beta_1 \neq 0)$.

b. The C.I. for β_1 is

$1.711 \pm (2.179)(.0997) = 1.711 \pm .217 = (1.494, 1.928)$. Thus the C.I. for $10\beta_1$ is $(14.94, 19.28)$.

Chapter 12: Regression and Correlation

39. We use the fact that $\hat{\beta}_1$ is unbiased for β_1. $E(\hat{\beta}_0) = \dfrac{E(\Sigma y_i - \hat{\beta}_1 \Sigma x_i)}{n}$

$$= \frac{E(\Sigma y_i)}{n} - E(\hat{\beta}_1)\bar{x} = \frac{E(\Sigma Y_i)}{n} - \beta_1\bar{x} = \frac{\Sigma(\beta_0 + \beta_1 x_i)}{n} - \beta_1\bar{x} = \beta_0 + \beta_1\bar{x} - \beta_1\bar{x} = \beta_0$$

41. $t = \hat{\beta}_1 \dfrac{\sqrt{\Sigma x_i^2 - (\Sigma x_i)^2/n}}{s}$. The numerator of $\hat{\beta}_1$ will be changed by the factor cd (since both $\Sigma x_i y_i$ and $(\Sigma x_i)(\Sigma y_i)$ appear) while the denominator of $\hat{\beta}_1$ will change by the factor c^2 (since both Σx_i^2 and $(\Sigma x_i)^2$ appear). Thus $\hat{\beta}_1$ will change by the factor d/c. Because $SSE = \Sigma(y_i - \hat{y}_i)^2$, SSE will change by the factor d^2, so s will change by the factor d. Since $\sqrt{\bullet}$ in t changes by the factor c, t itself will change by $\dfrac{d}{c}\cdot\dfrac{c}{d} = 1$, or not at all.

43. $H_0 : \beta_1 = 0$ vs $H_a : \beta_1 \neq 0$. The value of the test statistic is z = .73, with a corresponding p-value of .463. Since the p-value is greater than any sensible choice of alpha we do not reject H_0. There is insufficient evidence to claim that age has a significant impact on the presence of kyphosis.

45.

    a. The mean of the x data in Exercise 12.15 is $\bar{x} = 613.5$. Since x = 600 is closer to 613.5 than is x = 750, the quantity $(600 - \bar{x})^2$ must be smaller than $(750 - \bar{x})^2$. Therefore, since these quantities are the only ones that are different in the two $s_{\hat{y}}$ values, the $s_{\hat{y}}$ value for x = 600 must necessarily be smaller than the $s_{\hat{y}}$ for x = 750. Said briefly, the closer x is to $\bar{x}$, the smaller the value of $s_{\hat{y}}$.

    b. Error degrees of freedom $= n - 2 = 6$. $t_{.025,6} = 2.447$, so the interval estimate when x = 600 is : $2.723 \pm (2.447)(.190) = (2.258, 3.188)$.

    c. The 95% prediction interval is
$$\hat{y} \pm t_{.025,6}\sqrt{s^2 + s_{\hat{y}}^2} = 2.723 \pm (2.447)\sqrt{(.534)^2 + (.190)^2} = (1.336, 4.110).$$ Note that the prediction interval is much wider than the CI.

    d. For two 95% intervals, the simultaneous confidence level is at least $100(1 - 2(.05)) = 90\%$.

Chapter 12: Regression and Correlation

47.

a. A 95% CI for $\mu_{Y \cdot 500}$: $\hat{y}_{(500)} = -.311 + (.00143)(500) = .40$ and

$$s_{\hat{y}_{(500)}} = .131\sqrt{\frac{1}{13} + \frac{(500 - 471.54)^2}{131,519.23}} = .03775 \text{, so the interval is}$$

$$\hat{y}_{(500)} \pm t_{.025,11} \cdot s_{\hat{y}_{(500)}} = .40 \pm 2.210(.03775) = .40 \pm .08 = (.32,.48)$$

b. The width at x = 400 will be wider than that of x = 500 because x = 400 is farther away from the mean ($\bar{x} = 471.54$).

c. A 95% CI for β_1:

$$\hat{\beta}_1 \pm t_{.025,11} \cdot s_{\hat{\beta}_1} = .00143 \pm 2.201(.0003602) = (.000637,.002223)$$

d. We wish to test $H_0 : y_{(400)} = .25$ vs. $H_0 : y_{(400)} \neq .25$. The test statistic is

$$t = \frac{\hat{y}_{(400)} - .25}{s_{\hat{y}_{(400)}}} \text{, and we reject } H_0 \text{ if } |t| \geq t_{.025,11} = 2.201.$$

$$\hat{y}_{(400)} = -.311 + .00143(400) = .2614 \text{ and}$$

$$s_{\hat{y}_{(400)}} = .131\sqrt{\frac{1}{13} + \frac{(400 - 471.54)^2}{131,519.23}} = .0445 \text{, so the calculated}$$

$$t = \frac{.2614 - .25}{.0445} = .2561 \text{, which is not } \geq 2.201 \text{, so we do not reject } H_0. \text{ This}$$

sample data does not contradict the prior belief.

49. 95% CI: (462.1, 597.7); midpoint = 529.9; $t_{.025,8} = 2.306$;

$529.9 + (2.306)\left(\hat{s}_{\hat{\beta}_0 + \hat{\beta}_1(15)}\right) = 597.7$; $\hat{s}_{\hat{\beta}_0 + \hat{\beta}_1(15)} = 29.402$; 99% CI:

$529.9 \pm (3.355)(29.402) = (431.3, 628.5)$

Chapter 12: Regression and Correlation

51.

a. We wish to test $H_o : \beta_1 = 0$ vs $H_a : \beta_1 \neq 0$. The test statistic

$t = \dfrac{10.6026}{.9985} = 10.62$ leads to a p-value of $< .006$ ($2P(\,t > 4.0\,)$ from the 7 df row of table A.8), and H_o is rejected since the p-value is smaller than any reasonable α. The data suggests that this model does specify a useful relationship between chlorine flow and etch rate.

b. A 95% confidence interval for β_1: $10.6026 \pm (2.365)(.9985) = (8.24, 12.96)$. We can be highly confident that when the flow rate is increased by 1 SCCM, the associated expected change in etch rate will be between 824 and 1296 A/min.

c. A 95% CI for $\mu_{Y \cdot 3.0}$: $38.256 \pm 2.365 \left(2.546 \sqrt{\dfrac{1}{9} + \dfrac{9(3.0 - 2.667)^2}{58.50}} \right)$

$= 38.256 \pm 2.365(2.546)(.35805) = 38.256 \pm 2.156 = (36.100, 40.412)$, or 3610.0 to 4041.2 A/min.

d. The 95% PI is $38.256 \pm 2.365 \left(2.546 \sqrt{1 + \dfrac{1}{9} + \dfrac{9(3.0 - 2.667)^2}{58.50}} \right)$

$= 38.256 \pm 2.365(2.546)(1.06) = 38.256 \pm 6.398 = (31.859, 44.655)$, or 3185.9 to 4465.5 A/min.

e. The intervals for $x^* = 2.5$ will be narrower than those above because 2.5 is closer to the mean than is 3.0.

f. No. A value of 6.0 is not in the range of observed x values, therefore predicting at that point is meaningless.

g. Three 99% confidence intervals will have simultaneous confidence level of at least $(100 - 3(1))\% = 97\%$. $t_{.01,7} = 2.998$, and the resulting three 99% CIs, following the method of (c), are $(23.880, 34.128)$, $(29.929, 35.981)$, and $(35.067, 41.446)$.

Chapter 12: Regression and Correlation

53.

a. There is a linear pattern in the scatter plot, although the pot also shows a reasonable amount of variation about any straight line fit to the data. The simple linear regression model provides a sensible starting point for a formal analysis.

b. In testing $H_o : \beta_1 = 0$ vs $H_a : \beta_1 \neq 0$, Minitab calculates

$$t = \frac{-1.060}{.241} = -4.39 \text{ with a corresponding P-value of .001. Hence, } H_o \text{ is}$$

rejected. The simple linear regression model does appear to specify a useful relationship.

c. A confidence interval for $\beta_0 + \beta_1(75)$ is requested. The interval is centered at

$$\hat{\beta}_0 + \hat{\beta}_1(75) = 435.9 . \quad s_{\hat{\beta}_0 + \hat{\beta}_1(75)} = s\sqrt{\frac{1}{n} + \frac{n(75 - \bar{x})^2}{n\Sigma x_i^2 - (\Sigma x_i)^2}} = 14.83 \text{ (using } s =$$

54.803). Thus a 95% CI is $435.9 \pm (2.179)(14.83) = (403.6, 468.2)$.

55.
$$\hat{\beta}_0 + \hat{\beta}_1 x = \bar{Y} - \hat{\beta}_1\bar{x} + \hat{\beta}_1 x = \bar{Y} + (x - \bar{x})\hat{\beta}_1 = \frac{1}{n}\sum Y_i + \frac{(x - \bar{x})\sum(x_i - \bar{x})Y_i}{S_{XX}} = \sum d_i Y_i$$

where $d_i = \frac{1}{n} + \frac{(x - \bar{x})(x_i - \bar{x})}{S_{XX}}$. Thus, $Var(\hat{\beta}_0 + \hat{\beta}_1 x) = \sum d_i^2 Var(Y_i) = \sigma^2 \sum d_i^2$

$$= \sigma^2 \sum \left[\frac{1}{n^2} + 2\frac{(x - \bar{x})(x_i - \bar{x})}{nS_{XX}} + \frac{(x - \bar{x})^2(x_i - \bar{x})^2}{nS_{XX}^2} \right]$$

$$= \sigma^2 \left[n\frac{1}{n^2} + 2\frac{(x - \bar{x})\sum(x_i - \bar{x})}{nS_{XX}} + \frac{(x - \bar{x})^2\sum(x_i - \bar{x})^2}{S_{XX}^2} \right]$$

$$= \sigma^2 \left[\frac{1}{n} + 2\frac{(x - \bar{x}) \cdot 0}{nS_{XX}} + \frac{(x - \bar{x})^2 S_{XX}}{S_{XX}^2} \right] = \sigma^2 \left[\frac{1}{n} + \frac{(x - \bar{x})^2}{S_{XX}} \right]$$

Chapter 12: Regression and Correlation

57.

a. Summary values: $\Sigma x = 44{,}615$, $\Sigma x^2 = 170{,}355{,}425$, $\Sigma y = 3{,}860$,
$\Sigma y^2 = 1{,}284{,}450$, $\Sigma xy = 14{,}755{,}500$, $n = 12$. Using these values we calculate
$S_{xx} = 4{,}480{,}572.92$, $S_{yy} = 42{,}816.67$, and $S_{xy} = 404{,}391.67$. So

$$r = \frac{S_{xy}}{\sqrt{S_{xx}}\sqrt{S_{yy}}} = .9233.$$

b. The value of r does not depend on which of the two variables is labeled as the x variable. Thus, had we let x = RBOT time and y = TOST time, the value of r would have remained the same.

c. The value of r does no depend on the unit of measure for either variable. Thus, had we expressed RBOT time in hours instead of minutes, the value of r would have remained the same.

d. Both TOST time and ROBT time appear to have come from normally distributed populations.

Normal Probability Plot

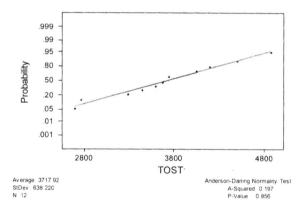

Average 3717 92
StDev 638 220
N 12

Anderson-Darling Normality Test
A-Squared 0 197
P-Value 0 856

216

Chapter 12: Regression and Correlation

Normal Probability Plot

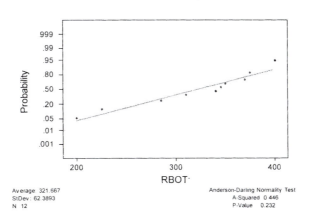

Average 321.667
StDev 62.3893
N 12

Anderson-Darling Normality Test
A-Squared 0.446
P-Value 0.232

e. $H_o : \rho_1 = 0$ vs $H_a : \rho \neq 0$. $t = \dfrac{r\sqrt{n-2}}{\sqrt{1-r^2}}$; Reject H_o at level .05 if either

$t \geq t_{.025,10} = 2.228$ or $t \leq -2.228$. r = .923, t = 7.58, so H_o should be rejected.
The model is useful.

59.

a. We are testing $H_o : \rho = 0$ vs $H_a : \rho > 0$.

$r = \dfrac{7577.632}{\sqrt{2628390.357}\sqrt{37.81269}} = .7600$, and $t = \dfrac{.762\sqrt{12}}{\sqrt{1-.76^2}} = 4.05$. We reject H_o

since $t = 4.05 \geq t_{.05,12} = 1.782$. There is evidence that a positive correlation
exists between maximum lactate level and muscular endurance.

b. We are looking for r^2, the coefficient of determination. $r^2 = (.76)^2 = 57.8\%$. It is
the same no matter which variable is the predictor.

Chapter 12: Regression and Correlation

61. $n = 6$, $\Sigma x_i = 111.71$, $\Sigma x_i^2 = 2{,}724.7643$, $\Sigma y_i = 2.9$, $\Sigma y_i^2 = 1.6572$, and $\Sigma x_i y_i = 63.915$.

$$r = \frac{(6)(63.915)-(111.71)(2.9)}{\sqrt{(6)(2{,}724.7943)-(111.73)^2} \cdot \sqrt{(6)(1.6572)-(2.9)^2}} = .7729 . \ H_o : \rho_1 = 0 \text{ vs}$$

$H_a : \rho \neq 0$; Reject H_o at level .05 if $|t| \geq t_{.025,4} = 2.776$. $\ t = \dfrac{(.7729)\sqrt{4}}{\sqrt{1-(.7729)^2}} = 2.436$.

Fail to reject H_o. The data does not indicate that the population correlation coefficient differs from 0. This result may seem surprising due to the relatively large size of r (.77), however, it can be attributed to a small sample size (6).

63.

    a.   Because p-value = .00032 < α = .001, H_o should be rejected at this significance level.

    b.   Not necessarily. For this n, the test statistic t has approximately a standard normal distribution when $H_o : \rho_1 = 0$ is true, and a p-value of .00032

        corresponds to $z = 3.60$ (or -3.60). Solving $3.60 = \dfrac{r\sqrt{498}}{\sqrt{1-r^2}}$ for r yields r =

        .159. This r suggests only a weak linear relationship between x and y, one that would typically have little practical importance.

    c. $t = \dfrac{.022\sqrt{9998}}{\sqrt{1-.022^2}} = 2.20 \geq t_{.025,9998} = 1.96$, so H_o is rejected in favor of H_a. The

        value t = 2.20 is statistically significant – it cannot be attributed just to sampling variability in the case $\rho = 0$. But with this n, r = .022 implies $\rho \approx .022$, which in turn shows an extremely weak linear relationship.

65. Re-write both statistics in terms of the original sums of squares. The test statistic from

Section 12.3 is $\dfrac{b_1 - 0}{S/\sqrt{S_{XX}}} = \dfrac{S_{XY}/S_{XX}}{\sqrt{SSE/n-2}/\sqrt{S_{XX}}} = \dfrac{S_{XY}\sqrt{n-2}}{\sqrt{SSE(S_{XX})}}$. Meanwhile, $T =$

$\dfrac{R\sqrt{n-2}}{\sqrt{1-R^2}} = \dfrac{(S_{XY}/\sqrt{S_{XX}S_{YY}})\sqrt{n-2}}{\sqrt{SSE/SST}} = \dfrac{S_{XY}\sqrt{n-2}}{\sqrt{SSE(S_{XX})}}$, since SST and S_{YY} are the

same thing.

67.

    a. We used Excel to calculate the r_i's: $r_1 = 0.184$, $r_2 = -0.238$, and $r_3 = -0.426$.

    b. The only difference between lag autocorrelation coefficients and regular correlation is the number of terms in the numerator summand: the sum only runs from 1 to $n - 1$, but $\bar{x}$ is based upon all n observations. In regular correlation, $\bar{x}$ would be replaced in each part of the numerator by the mean of just the relevant $n - 1$ values (1 through $n - 1$ in the first parentheses, 2 through n in the second). As n gets larger, the difference between $\bar{x}$ and these "truncated" means becomes negligible. A similar comment applies to lag 2.

    c. $\dfrac{2}{\sqrt{100}} = .2$. We reject H_o if $|r_i| \geq .2$. For all lags, r_i does not fall in the rejection region, so we cannot reject H_o. There is not evidence of theoretical autocorrelation at the first 3 lags.

    d. If we want an approximate .05 significance level for the simultaneous hypotheses, we would have to use smaller individual significance level. If the individual confidence levels were .95, then the simultaneous confidence levels would be approximately $(.95)(.95)(.95) = .857$.

69.   The pattern gives no cause for questioning the appropriateness of the simple linear regression model, and no observation appears unusual.

Chapter 12: Regression and Correlation

71.

a. The (x, residual) pairs for the plot are (0, -.335), (7, -.508), (17. -.341), (114, .592), (133, .679), (142, .700), (190, .142), (218, 1.051), (237, -1.262), and (285, -.719). The plot shows substantial evidence of curvature.

b. The standardized residuals (in order corresponding to increasing x) are -.50, -.75, -.50, .79, .90, .93, .19, 1.46, -1.80, and -1.12. A standardized residual plot shows the same pattern as the residual plot discussed in the previous exercise. The z percentiles for the normal probability plot are –1.645, -1.04, -.68, -.39, -.13, .13, .39, .68, 1.04, 1.645. The plot follows. The points follow a linear pattern, so the standardized residuals appear to have a normal distribution.

Normal Probability Plot for the Standardized Residuals

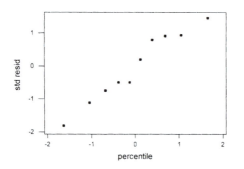

73.

a. $H_o : \beta_1 = 0$ vs. $H_a : \beta_1 \neq 0$. The test statistic is $t = \dfrac{\hat{\beta}_1}{s_{\hat{\beta}_1}}$, and we will reject

H$_o$ if $t \geq t_{.025,4} = 2.776$ or if $t \leq -2.776$. $s_{\hat{\beta}_1} = \dfrac{s}{\sqrt{S_{xx}}} = \dfrac{7.265}{12.869} = .565$, and

$t = \dfrac{6.19268}{.565} = 10.97$. Since $10.97 \geq 2.776$, we are tempted to reject H$_o$ and

conclude that the model is useful. However, this test assumes that a true linear relationship exists between x and y, which is contradicted by the residual plots below.

b. $\hat{y}_{(7.0)} = 1008.14 + 6.19268(7.0) = 1051.49$, from which the residual is

$y - \hat{y}_{(7.0)} = 1046 - 1051.49 = -5.49$. Similarly, the other residuals are -.73, 4.11, 7.91, 3.58, and –9.38. The plot of the residuals vs x follows:

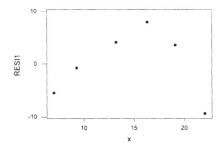

Because a curved pattern appears, a linear regression function is inadequate.

c. The standardized residuals are calculated as

$$e_1{}^* = \frac{-5.49}{7.265\sqrt{1 + \dfrac{1}{6} + \dfrac{(7.0 - 14.48)^2}{165.5983}}} = -1.074 \text{, and similarly the others}$$

are -.123, .624, 1.208, .587, and –1.841. The plot of e* vs x follows :

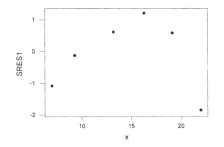

This plot gives the same information as the previous plot. No values are exceptionally large, but the e* of –1.841 is close to 2 std deviations away from the expected value of 0.

Chapter 12: Regression and Correlation

75. Both a scatter plot and residual plot (based on the simple linear regression model) for the first data set suggest that a simple linear regression model is reasonable, with no pattern or influential data points which would indicate that the model should be modified. However, scatter plots for the other three data sets reveal difficulties.

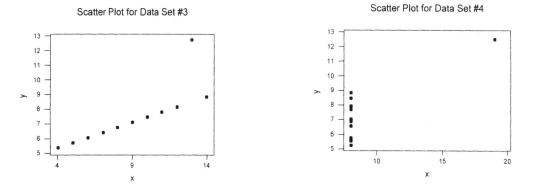

Scatter Plot for Data Set #3

Scatter Plot for Data Set #4

For data set #2, a quadratic function would clearly provide a much better fit. For data set #3, the relationship is perfectly linear except one outlier, which has obviously greatly influenced the fit even though its x value is not unusually large or small. One might investigate this observation to see whether it was mistyped and/or it merits deletion. For data set #4 it is clear that the slope of the least squares line has been determined entirely by the outlier, so this point is extremely influential. A linear model is completely inappropriate for data set #4; if anything, a modified logistic regression model (see Section 13.2) might be more appropriate.

Chapter 12: Regression and Correlation

77.

a. $n_1 = n_2 = 3$ (3 observations at 110 and 3 at 230), $n_3 = n_4 = 4$,

$\bar{y}_{1.} = 202.0$, $\bar{y}_{2.} = 149.0$, $\bar{y}_{3.} = 110.5$, $\bar{y}_{4.} = 107.0$, $\Sigma\Sigma y_{ij}^2 = 288{,}013$,

so $SSPE = 288{,}013 - \left[3(202.0)^2 + 3(149.0)^2 + 4(110.5)^2 + 4(107.0)^2\right] = 4361$.

With $\Sigma x_i = 4480$, $\Sigma y_i = 1923$, $\Sigma x_i^2 = 1{,}733{,}500$, $\Sigma y_i^2 = 288{,}013$ (as

above), and $\Sigma x_i\, y_i = 544{,}730$, SSE = 7241 so SSLF = 7241-4361=2880.

With c – 2 = 2 and n – c = 10, $F_{.05,2,10} = 4.10$. $MSLF = \dfrac{2880}{2} = 1440$ and

$SSPE = \dfrac{4361}{10} = 436.1$, so the computed value of F is $\dfrac{1440}{436.1} = 3.30$. Since

3.30 is not ≥ 4.10, we do not reject H_0. This formal test procedure does not suggest that a linear model is inappropriate.

b. The scatter plot clearly reveals a curved pattern which suggests that a nonlinear model would be more reasonable and provide a better fit than a linear model.

79.

a. For $x_1 = 2$, $x_2 = 8$ (remember the units of x_2 are in 1000s) and $x_3 = 1$ (since the outlet has a drive-up window) the average sales are
$\hat{y} = 10.00 - 1.2(2) + 6.8(8) + 15.3(1) = 77.3$ (i.e., \$77,300).

b. For $x_1 = 3$, $x_2 = 5$, and $x_3 = 0$ the average sales are
$\hat{y} = 10.00 - 1.2(3) + 6.8(5) + 15.3(0) = 40.4$ (i.e., \$40,400).

c. When the number of competing outlets (x_1) and the number of people within a 1-mile radius (x_2) remain fixed, the expected sales will increase by \$15,300 when an outlet has a drive-up window.

81. $H_0 : \beta_1 = \beta_2 = \ldots = \beta_6 = 0$ vs. H_a: at least one among $\beta_1, \ldots, \beta_6$ is not zero. The test

statistic is $F = \dfrac{R^2/k}{(1-R^2)/(n-k-1)}$. H_0 will be rejected if $f \geq F_{.05,6,30} = 2.42$.

$f = \dfrac{.83/6}{(1-.83)/30} = 24.41$. Because $24.41 \geq 2.42$, H_0 is rejected and the model is judged

useful.

Chapter 12: Regression and Correlation

83.

a. $x_1 = 2.6$, $x_2 = 250$, and $x_1x_2 = (2.6)(250) = 650$, so
$\hat{y} = 185.49 - 45.97(2.6) - 0.3015(250) + 0.0888(650) = 48.313$

b. No, it is not legitimate to interpret β_1 in this way. It is not possible to increase by 1 unit the cobalt content, x_1, while keeping the interaction predictor, x_3, fixed. When x_1 changes, so does x_3, since $x_3 = x_1x_2$.

c. Yes, there appears to be a useful linear relationship between y and the predictors. We determine this by observing that the p-value corresponding to the model utility test is $< .0001$ (F test statistic = 18.924).

d. We wish to test $H_0 : \beta_3 = 0$ vs. $H_a : \beta_3 \neq 0$. The test statistic is t=3.496, with a corresponding p-value of .0030. Since the p-value is $<$ alpha = .01, we reject H_0 and conclude that the interaction predictor does provide useful information about y.

e. A 95% C.I. for the mean value of surface area under the stated circumstances requires the following quantities:
$\hat{y} = 185.49 - 45.97(2) - 0.3015(500) + 0.0888(2)(500) = 31.598$. Next,
$t_{.025,16} = 2.120$, so the 95% confidence interval is
$31.598 \pm (2.120)(4.69) = 31.598 \pm 9.9428 = (21.6552, 41.5408)$

f. There appear to be no problems with normality or curvature, but the variance may depend on x_1.

85.

a. No, there is no pattern in the plots which would indicate that a transformation or the inclusion of other terms in the model would produce a substantially better fit.

b. k = 5, n – (k+1) = 8, so $H_0 : \beta_1 = \ldots = \beta_5 = 0$ will be rejected if
$f \geq F_{.05,5,8} = 3.69$; $f = \dfrac{(.759)/5}{(.241)/8} = 5.04 \geq 3.69$, so we reject H_0. At least one of
the coefficients is not equal to zero.

c. This is unusual, but not impossible: a global F test of all 5 coefficients detects significant contribution from the explanatory variables, but 5 separate t tests do not. This is analogous to the unusual (but, again, not impossible) situation in ANOVA where the F test indicates not all means are equal, but Tukey's HSD method does not find any statistically significant differences.

Chapter 12: Regression and Correlation

d. When $x_1 = 8.0$ and $x_2 = 33.1$ the residual is e = 2.71 and the standardized residual is e* = .44; since e* = e/(sd of the residual), sd of residual = e/e* = 6.16. Thus the estimated variance of $\hat{Y}$ is $(6.99)^2 - (6.16)^2 = 10.915$, so the estimated sd is 3.304. Since $\hat{y} = 24.29$ and $t_{.025,8} = 2.306$, the desired C.I. is
$24.29 \pm 2.306(3.304) = (16.67, 31.91)$.

e. $F_{.05,3,8} = 4.07$, so $H_0 : \beta_3 = \beta_4 = \beta_5 = 0$ will be rejected if $f \geq 4.07$. With
$SSE_k = 8, s^2 = 390.88$, and $f = \dfrac{(894.95 - 390.88)/3}{(390.88)/8} = 3.44$, and since 3.44 is not

≥ 4.07, H_o cannot be rejected and the quadratic terms should all be deleted. (n.b.: this is not a modification which would be suggested by a residual plot)

87.

a. The complete 2nd order model obviously provides a better fit, so there is a need to account for interaction between the three predictors.

b. From Minitab, s = .044. So, a 95% PI for y when $x_1 = x_2 = 30$ and $x_3 = 10$ is
$.66573 \pm 2.228\sqrt{.044^2 + .01785^2} = (.560, .771)$.

89.

a. The data and response matrices are $\mathbf{X} = \begin{bmatrix} 1 & -1 & -1 \\ 1 & -1 & 1 \\ 1 & 1 & -1 \\ 1 & 1 & 1 \end{bmatrix}$ and $\mathbf{y} = \begin{bmatrix} 1 \\ 1 \\ 0 \\ 4 \end{bmatrix}$. The normal equations are $\mathbf{X'Xb} = \mathbf{X'y}$, which here become $\begin{bmatrix} 4 & 0 & 0 \\ 0 & 4 & 0 \\ 0 & 0 & 4 \end{bmatrix} \mathbf{b} = \begin{bmatrix} 6 \\ 2 \\ 4 \end{bmatrix}$.

b. Since $\mathbf{X'X} = 4\mathbf{I}$, $(\mathbf{X'X})^{-1} = .25\mathbf{I}$, and $\mathbf{b} = .25\mathbf{X'y} = \begin{bmatrix} 1.5 \\ 0.5 \\ 1.0 \end{bmatrix}$.

c. $\hat{\mathbf{y}} = \mathbf{Xb} = \begin{bmatrix} 0 \\ 2 \\ 1 \\ 3 \end{bmatrix}$, from which SSE $= \|\mathbf{y} - \hat{\mathbf{y}}\|^2 = (1)^2 + (-1)^2 + (-1)^2 + (1)^2 = 4$, and
MSE = SSE/[4-(2+1)] = SSE/1 = 4.

Chapter 12: Regression and Correlation

d. A 95% CI for β_1 is $b_1 \pm t_{.025,4-3}s\sqrt{c_{11}}$. From above, $b_1 = 0.5$, $s^2 = \text{MSE} = 4$, and $c_{11} = .25$; $t_{.025,1} = 12.706$. The resulting CI is $0.5 \pm 12.706 = (-12.206, 13.206)$. The CI is so large because we only have 1 df (4 observations, 3 parameters).

e. The t statistic here is $t = \dfrac{0.5 - 0}{\sqrt{4}\sqrt{.25}} = 0.5$, which at 1df has a 2-sided P-value of $2(.352) = .704$. We certainly fail to reject the hypothesis that $\beta_1 = 0$. This is consistent with our 95% CI from part (d).

f. $\bar{y} = 6/4 = 1.5$, so $\text{SSR} = \|\hat{\mathbf{y}} - \bar{\mathbf{y}}\|^2 = (-1.5)^2 + (.5)^2 + (-.5)^2 + (1.5)^2 = 5$. The rest of the ANOVA table below (supplied by Minitab) follows. In particular, the F test statistic is $f = 0.63$ with a corresponding P-value of .667, so we definitely fail to reject H_0. Both slopes could plausibly be zero, and so it appears neither x_1 nor x_2 is a useful predictor for y. Finally, $R^2 = \text{SSR}/\text{SST} = 5/9 = 55.56\%$; that is, ~56% of the variability in y can be explained by the linear regression model that involves predictors x_1 and x_2.

```
Source            DF    SS      MS      F      P
Regression         2   5.000   2.500   0.63   0.667
Residual Error     1   4.000   4.000
Total              3   9.000
```

91. In this case, $\mathbf{X}$ is an $n \times 1$ vector of 1's, so $\mathbf{X'X} = [n]$ and $\mathbf{X'y} = [\sum y_i]$. Thus, $[b_0] = (\mathbf{X'X})^{-1}\mathbf{X'y} = [\bar{y}]$. Hence, $\hat{\mathbf{y}}$ is an $n \times 1$ vectors whose every entry is $\bar{y}$, and $(n-1)s^2 = \text{MSE} = \|\mathbf{y} - \hat{\mathbf{y}}\|^2 = \sum y_i - \bar{y})^2$, or $s = s_y$. Lastly, $c_{00} = 1/n$, and so a 95% CI for β_0 is
$$b_0 \pm t_{.025,n-(0-1)}s\sqrt{c_{00}} = \bar{y} \pm t_{.025,n-1}s/\sqrt{n}\,,\text{ our usual 95% CI for } \mu_y.$$

93.

a. With the given x-values, $\mathbf{X'X} = \begin{bmatrix} m+n & \sum x_i \\ \sum x_i & \sum x_i^2 \end{bmatrix} = \begin{bmatrix} 2n & 0 \\ 0 & n/2 \end{bmatrix}$. Also, $\mathbf{X'y} = $

$\begin{bmatrix} \sum_{i=1}^{n+m} y_i \\ .5\sum_{i=1}^{m} y_i - .5\sum_{i=m+1}^{m+n} y_i \end{bmatrix} = \begin{bmatrix} n(\bar{y}_1 + \bar{y}_2) \\ n(\bar{y}_1 - \bar{y}_2)/2 \end{bmatrix}$, from which $\mathbf{b} = (\mathbf{X'X})^{-1}\mathbf{X'y} = $

$\begin{bmatrix} 1/2n & 0 \\ 0 & 2/n \end{bmatrix}\begin{bmatrix} n(\bar{y}_1 + \bar{y}_2) \\ n(\bar{y}_1 - \bar{y}_2)/2 \end{bmatrix} = \begin{bmatrix} (\bar{y}_1 + \bar{y}_2)/2 \\ \bar{y}_1 - \bar{y}_2 \end{bmatrix}$. That is, $b_0 = (\bar{y}_1 + \bar{y}_2)/2$ and $b_1 = \bar{y}_1 - \bar{y}_2$.

b. From above, $c_{11} = 2/n$. $\hat{\mathbf{y}} = \mathbf{Xb} = [\bar{y}_1 \; \dots \; \bar{y}_1 \; \bar{y}_2 \; \dots \; \bar{y}_2]'$, so $\text{SSE} = \|\mathbf{y} - \hat{\mathbf{y}}\|^2 = \sum_{i=1}^{m}(y_i - \bar{y}_1)^2 + \sum_{i=m+1}^{m+n}(y_i - \bar{y}_2)^2$ and $s^2 = \text{MSE} = \text{SSE}/(m+n-2)$.

c. A 95% CI for β_1 is $b_1 \pm t_{.025, m+n-2} s \sqrt{c_{11}}$ $= (\bar{y}_1 - \bar{y}_2) \pm t_{.025, m+n-2}$

$$\sqrt{\frac{\sum_{i=1}^{m}(y_i - \bar{y}_1)^2 + \sum_{i=m+1}^{m+n}(y_i - \bar{y}_2)^2}{m+n-2}} \sqrt{2/n} \text{ ; this is a match to the suggested CI,}$$

once we note that $1/m + 1/n = 2/n$ for the last term.

d. With $\mathbf{X'} = \begin{vmatrix} 1 & 1 & 1 & 1 & 1 & 1 \\ .5 & .5 & .5 & -.5 & -.5 & -.5 \end{vmatrix}$ and $\mathbf{y'} = [117\ 119\ 127\ 129\ 138\ 139]$, we get the

following: $\mathbf{b'} = [128.166, -14.333]$; $\hat{\mathbf{y}}' = [121\ 121\ 121\ 135.33\ 135.33\ 135.33]$;

SSE $= \ldots = 116.666$, $s = 5.4$, $c_{11} = 2/3$. Finally, the 95% CI for β_1 is $-14.333 \pm$

$2.776(5.4)\sqrt{2/3} = (-26.58, -2.09)$.

95. The Residual column is, obviously, just the difference of the first two columns
(observed – predicted). The numbers under Std Error Residual are the square roots of
the diagonal entries of the matrix $s^2(I - H)$. These diagonal entries are $s^2(I_{jj} - H_{jj}) = s^2$
$- s^2 H_{jj}$. If we have the ANOVA table, we can find $s^2 = $ MSE; the value $s^2 H_{jj}$ is just the
square of the entry under Std Error Mean Predict. That is, Std Error Residual $=$

$\sqrt{\text{MSE} - \text{SEMP}^2}$. The last column, Student Residual, is just the ratio of the previous
two.

97. Focus on the diagonal elements of these two covariance matrices: The ith diagonal
entry of $\sigma^2(I - H)$ is $V(y_i - \hat{y}_i)$, but it's also $\sigma^2(1 - h_{ii})$. Since variances can never be
negative, this means $\sigma^2(1 - h_{ii}) \geq 0$, or $h_{ii} \leq 1$. Similarly, $V(\hat{y}_i) = $ the ith diagonal
entry of $\sigma^2 H = \sigma^2 h_{ii}$ implies $h_{ii} \geq 0$. Hence, these exact same variances are between 0
and σ^2.

99.

a. From Exercise 92, $(\mathbf{X'X})^{-1}\mathbf{X'} = \begin{bmatrix} 1/n & 0 \\ 0 & 1/S_{XX} \end{bmatrix} \begin{bmatrix} 1 & \ldots & 1 \\ x_1 - \bar{x} & \ldots & x_n - \bar{x} \end{bmatrix} =$

$\begin{bmatrix} 1/n & \ldots & 1/n \\ (x_1 - \bar{x})/S_{XX} & \ldots & (x_n - \bar{x})/S_{XX} \end{bmatrix}$. Then, $\mathbf{H} = \mathbf{X}(\mathbf{X'X})^{-1}\mathbf{X'}$, an n-by-n matrix

with entries $h_{ij} = \frac{1}{n} + \frac{(x_i - \bar{x})(x_j - \bar{x})}{S_{XX}}$. In particular,

$V(\hat{y}_i) = \sigma^2 h_{ii} = \sigma^2 \left[\frac{1}{n} + \frac{(x_i - \bar{x})^2}{S_{XX}} \right]$.

b. As noted in the text, $\text{cov}(\mathbf{Y} - \hat{\mathbf{Y}}) = \sigma^2(\mathbf{I} - \mathbf{H})$, so the variance of the ith residual is

$$\sigma^2(1 - h_{ii}) = \sigma^2 - \sigma^2 h_{ii} = \sigma^2\left[1 - \frac{1}{n} - \frac{(x_i - \bar{x})^2}{S_{XX}}\right].$$

c. Consider the expression in (a). The further x is from $\bar{x}$, the greater the second term in $V(\hat{y}_i)$ will be.

d. Consider the expression in (b). The further x is from $\bar{x}$, the greater the third term will be, resulting in a <u>lower</u> variance for the residual.

e. Part (c) is, in a sense, the issue of extrapolation: the further x is from $\bar{x}$, the less reliable (less precise) our predictions are. On the other hand, the further x is from $\bar{x}$, the greater its influence over the regression line. Hence, x-values far away from the mean will consistently draw the regression line towards themselves, making the corresponding residuals smaller. Observations with x-values near the mean don't have nearly as much influence on the line, so the residuals at those values are more disperse.

101.

a. Using the results in this section, the vector of coefficients is $\mathbf{b'} = [35.0 \ 3.18 - .006]$. The model utility test gives $f = 12.04$ (P-value $< .001$); the t test for β_1 (foot) is $t = 2.96$ (P-value $= .021$ from Minitab); the t test for β_2 (height) is $t = -0.02$ (P-value $= .981$ from Minitab). That is, the overall model is useful for predicting wingspan, and foot size is a useful predictor. However, in the presence of foot size, height is a basically useless addition to the model.

b. The diagonal entries of $\mathbf{H}$, in order, are: .55, .31, .13, .11, .88, .17, .31, .15, .18, .20. Observation #5 has the highest leverage by far, by grace of the fact that the height (54") is much lower than any other observed height. 54" is 4'6", suggesting that student #5 mis-recorded his own height (perhaps it should be 64"). It's also hard to believe that a 4'6" person would wear a size 9 shoe.

c. Students #1 and #7 ($h = .55, .31$) are very tall and have very big feet. Student #2 has rather small feet, both for his height and for the group overall.

d. Student #2 has a very large, negative residual. It seems that a 56" wing span for a 66" person is rather short.

e. If numbers were clearly mis-recorded, these observations should be corrected or deleted. In general, though, we do not delete a "correct" observation simply because it doesn't follow the pattern suggested by the other observations.

Chapter 12: Regression and Correlation

103.

  a. $r^2 = .5073$

  b. $r = +\sqrt{r^2} = \sqrt{.5073} = .7122$ (positive because $\hat{\beta}_1$ is positive.)

  c. We test test $H_0 : \beta_1 = 0$ vs $H_0 : \beta_1 \neq 0$. The test statistic t = 3.93 gives p-value = .0013, which is < .01, the given level of significance, therefore we reject H_0 and conclude that the model is useful.

  d. We use a 95% CI for $\mu_{Y \cdot 50}$. $\hat{y}_{(50)} = .787218 + .007570(50) = 1.165718$, $t_{.025,15} = 2.131$, s = "Root MSE" = .020308, so

$$s_{\hat{y}_{(50)}} = .20308\sqrt{\frac{1}{17} + \frac{17(50-42.33)^2}{17(41,575)-(719.60)^2}} = .051422. \text{ The interval is, then,}$$

$$1.165718 \pm 2.131(.051422) = 1.165718 \pm .109581 = (1.056137, 1.275299).$$

  e. $\hat{y}_{(30)} = .787218 + .007570(30) = 1.0143$. The residual is $y - \hat{y} = .80 - 1.0143 = -.2143$.

105. Substituting x* = 0 gives the CI $\hat{\beta}_0 \pm t_{\alpha/2.n-2} \cdot s\sqrt{\frac{1}{n} + \frac{n\bar{x}^2}{n\Sigma x_i^2 - (\Sigma x_i)^2}}$. From

Example 12.8, $\hat{\beta}_0 = 3.621$, SSE = .262453, n = 14,

$\Sigma x_i = 890, \bar{x} = 63.5714, \Sigma x_i^2 = 67,182$, so with s = .1479, $t_{.025,12} = 2.179$, the CI is

$$3.621 \pm 2.179(.1479)\sqrt{\frac{1}{12} + \frac{56,578.52}{148,448}}$$

$$= 3.621 \pm 2.179(.1479)(.6815) = 3.62 \pm .22 = (3.40, 3.84).$$

107. The value of the sample correlation coefficient using the squared y values would not necessarily be approximately 1. If the y values are greater than 1, then the squared y values would differ from each other by more than the y values differ from one another. Hence, the relationship between x and y^2 would be less like a straight line, and the resulting value of the correlation coefficient would decrease.

Chapter 12: Regression and Correlation

109.

a. A scatter plot suggests the linear model is appropriate.

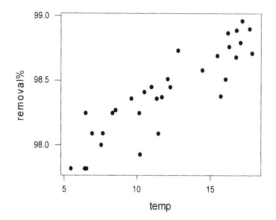

b. Minitab Output:
```
The regression equation is
removal% = 97.5 + 0.0757 temp

Predictor        Coef        StDev          T          P
Constant      97.4986       0.0889    1096.17      0.000
temp         0.075691     0.007046      10.74      0.000

S = 0.1552      R-Sq = 79.4%      R-Sq(adj) = 78.7%

Analysis of Variance

Source            DF          SS          MS          F          P
Regression         1      2.7786      2.7786     115.40      0.000
Residual Error    30      0.7224      0.0241
Total             31      3.5010
```

Minitab will output all the residual information if the option is chosen, from which you can find the point prediction value $\hat{y}_{10.5} = 98.2933$, the observed value $y = 98.41$, so the residual = .0294.

c. Roughly $s = .1552$

d. $R^2 = 79.4$

e. A 95% CI for β_1, using $t_{.025,30} = 2.042$:

$.075691 \pm 2.042(.007046) = (.061303, .090079)$

f. The slope of the regression line is steeper. The value of s is almost doubled (to 0.291), and the value of R^2 drops to 61.6%.

Chapter 12: Regression and Correlation

111.

a. Using the techniques from a previous chapter, we can do a t test for the difference of two means based on paired data. Minitab's paired t test for equality of means gives t = 3.54, with a p value of .002, which suggests that the average bf% reading for the two methods is not the same.

b. Using linear regression to predict HW from BOD POD seems reasonable after looking at the scatterplot, below.

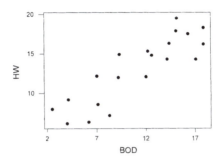

The least squares linear regression equation, as well as the test statistic and p value for a model utility test, can be found in the Minitab output below. We see that we do have significance, and the coefficient of determination shows that about 75% of the variation in HW can be explained by the variation in BOD.

```
The regression equation is
HW = 4.79 + 0.743 BOD

Predictor        Coef        StDev           T          P
Constant        4.788        1.215        3.94      0.001
BOD            0.7432       0.1003        7.41      0.000

S = 2.146      R-Sq = 75.3%      R-Sq(adj) = 73.9%

Analysis of Variance

Source          DF          SS          MS          F          P
Regression       1       252.98      252.98      54.94      0.000
Residual Error  18        82.89        4.60
Total           19       335.87
```

231

Chapter 12: Regression and Correlation

113. Use what we already know about MLE's of normal random samples. In the unrestricted case, $\hat{\sigma}^2 = \frac{1}{n}\Sigma(y_i - \hat{\mu})^2 = \frac{1}{n}\Sigma(y_i - b_0 - b_1 x_i)^2 = \frac{1}{n}\Sigma(y_i - \hat{y}_i)^2$. Under $H_0: \beta_1 = 0$, $\mu = \beta_0$, so $\hat{\beta}_0 = \bar{y}$ and

$\hat{\sigma}_0^2 = \frac{1}{n}\Sigma(y_i - \hat{\mu})^2 = \frac{1}{n}\Sigma(y_i - \hat{\beta}_0 - 0x_i)^2 = \frac{1}{n}\Sigma(y_i - \bar{y})^2$. Lastly, the exponential terms in the likelihood simplify to exp($-n/2$) in both cases, for a likelihood ratio equal to $\dfrac{(2\pi\hat{\sigma}_0^2)^{-n/2}\exp(-n/2)}{(2\pi\hat{\sigma}^2)^{-n/2}\exp(-n/2)} = \left(\dfrac{\hat{\sigma}^2}{\hat{\sigma}_0^2}\right)^{n/2} = \left(\dfrac{SSR}{SST}\right)^{n/2}$. We reject H_0 when this ratio is small, which (by the ANOVA equation) is equivalent to SSR/SSE being large, or $F =$ MSR/MSE being large.

115.

   a. Take logs of both sides of the model to get $\ln(Y) = \ln(\alpha) + \beta x + \ln(\varepsilon)$, or $Y' = \beta_0 + \beta_1 x + \varepsilon'$. If ε is lognormal, then (by definition) ε' is normal, and we have our usual regression model.

   b. Minitab gives the estimated regression equation $y = 299 - 3.46x$. However, residual plots show a strong curved patter among the residuals, and that the residuals are non-normal. The former indicates the simple linear model is <u>not</u> adequate.

   c. A scatter plot (y vs. x) does show a pattern consistent with a (decreasing) exponential model. And a scatter plot of $\ln(y)$ versus x is quite linear. If we regress $\ln(y)$ on x, Minitab gives the estimated regression equation $\ln(y) = 6.02 - 0.0333x$. Minitab also gives $r^2 = 98.8\%$, a good sign of fit, and residual plots don't show any general pattern (a good sign for model adequacy). The estimates of the original parameters are $a = \exp(b_0) = 411.6$ and $b = b_1 = -0.0333$.

   d. From Minitab, a 95% PI for $\ln(y)$ when x = 50 is (4.0630,4.6461). Thus, a 95% PI for y when x = 50 is $(\exp(4.0630),\exp(4.6461)) = (58.15,104.18)$.

Chapter 12: Regression and Correlation

117.

a. The scatter plot below strongly suggests a quadratic model.

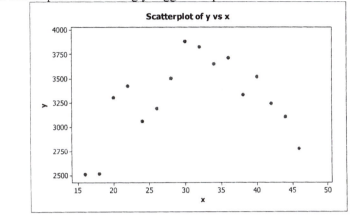

b. Minitab gives the estimated regression function $y = -1070.4 + 293.48x - 4.5358x^2$. The model utility F test has $f = 25.08$ (P-value = 0.000), so the model is quite useful.

c. From Minitab, a 95% CI for y when x = 25 is (3282.2,3581.4), while the corresponding 95% PI is (2966.5,3897.1). As usual, the PI is significantly wider than the CI.

d. From Minitab, a 95% CI for y when x = 40 is (3257.6,3565.6), while the corresponding 95% PI is (2945.0,3878.2). These are roughly the same width as the intervals in (c). That's not too surprising, since 40 is not too much further from the "middle" of the x-values than 25.

e. From Minitab, the t statistic for testing $H_0: \beta_2 = 0$ is -6.73 with a P-value of 0.000. Hence, we strongly reject this hypothesis; the quadratic term definitely provides useful information for predicting y.

119.

    a. 80.79

    b. Yes, p-value = .007 which is less than .01.

    c. No, p-value = .043 which is less than .05.

    d. $.14167 \pm (2.447)(.03301) = (.0609, .2224)$

    e. $\hat{\mu}_{y \cdot 9.66} = 6.3067$, using $\alpha = .05$, the interval is

$$6.3067 \pm (2.447)\sqrt{(.4851)^2 + (.162)^2} = (5.06, 7.56)$$

Chapter 13: Goodness-of-Fit Tests and Categorical Data Analysis

1.

  a.   We reject H_o if the calculated χ^2 value is greater than or equal to the tabled value

   of $\chi^2_{\alpha,k-1}$ from Table A.7. Since $12.25 \geq \chi^2_{.05,4} = 9.488$, we would reject H_o.

  b.   Since 8.54 is not $\geq \chi^2_{.01,3} = 11.344$, we would fail to reject H_o.

  c.   Since 4.36 is not $\geq \chi^2_{.10,2} = 4.605$, we would fail to reject H_o.

  d.   Since 10.20 is not $\geq \chi^2_{.01,5} = 15.085$, we would fail to reject H_o.

3.   Using the number 1 for business, 2 for engineering, 3 for social science, and 4 for
agriculture, let p_i = the true proportion of all clients from discipline i. If the
Statistics department's expectations are correct, then the relevant null hypothesis is
$H_o : p_1 = .40, p_2 = .30, p_3 = .20, p_4 = .10$, versus H_a : The Statistics department's
expectations are not correct. With d.f = k – 1 = 4 – 1 = 3, we reject H_o if
$\chi^2 \geq \chi^2_{.05,3} = 7.815$. Using the proportions in H_o, the expected number of clients are:

Client's Discipline	Expected Number
Business	(120)(.40) = 48
Engineering	(120)(.30) = 36
Social Science	(120)(.20) = 24
Agriculture	(120)(.10) = 12

Since all the expected counts are at least 5, the chi-squared test can be used. The

value of the test statistic is $\chi^2 = \displaystyle\sum_{i=1}^{k} \frac{(n_i - np_i)^2}{np_i} = \sum_{allcells} \frac{(observed - expected)^2}{expected}$

$= \left[\dfrac{(52-48)^2}{48} + \dfrac{(38-36)^2}{36} + \dfrac{(21-24)^2}{24} + \dfrac{(9-12)^2}{12} \right] = 1.57$, which is not ≥ 7.815, so

we fail to reject H_o. (Alternatively, p-value = $P(\chi^2 \geq 1.57)$ which is > .10, and since
the p-value is not < .05, we reject H_o). Thus we have no significant evidence to
suggest that the statistics department's expectations are incorrect.

Chapter 13: Goodness-of-Fit Tests and Categorical Data Analysis

5. We will reject H_o if the p-value $< .10$. The observed values, expected values, and corresponding χ^2 terms are:

Obs	4	15	23	25	38	21	32	14	10	8
Exp	6.67	13.33	20	26.67	33.33	33.33	26.67	20	13.33	6.67
χ^2	1.069	.209	.450	.105	.654	.163	1.065	1.800	.832	.265

$\chi^2 = 1.069 + ... + .265 = 6.612$. With d.f. $= 10 - 1 = 9$, our χ^2 value of 6.612 is less than $\chi^2_{.10,9} = 14.684$, so the p-value $> .10$ and we cannot reject H_o. There is no significant evidence that the data is not consistent with the previously determined proportions.

7. We test $H_o : p_1 = p_2 = p_3 = p_4 = .25$ vs. H_a : at least one proportion $\neq .25$, and d.f. $= 3$. We will reject H_o if the p-value $< .01$.

Cell	1	2	3	4
Observed	328	334	372	327
Expected	340.25	340.25	340.25	34.025
χ^2 term	.4410	.1148	2.9627	.5160

$\chi^2 = 4.0345$, and with 3 d.f., p-value $> .10$, so we fail to reject H_o. The data fails to indicate a seasonal relationship with incidence of violent crime.

9.

a. Denoting the 5 intervals by $[0, c_1), [c_1, c_2), ..., [c_4, \infty)$, we wish c_1 for which
$$.2 = P(0 \le X \le c_1) = \int_0^{c_1} e^{-x} dx = 1 - e^{-c_1}, \text{ so } c_1 = -\ln(.8) = .2231. \text{ Then}$$
$.2 = P(c_1 \le X \le c_2) \Rightarrow .4 = P(0 \le X_1 \le c_2) = 1 - e^{-c_2}, \text{ so } c_2 = -\ln(.6) = .5108.$
Similarly, $c_3 = -\ln(.4) = .0163$ and $c_4 = -\ln(.2) = 1.6094$. the resulting intervals are $[0, .2231), [.2231, .5108), [.5108, .9163), [.9163, 1.6094), \text{ and } [1.6094, \infty)$.

b. Each expected cell count is $40(.2) = 8$, and the observed cell counts are 6, 8, 10, 7, and 9, so $\chi^2 = \left[\dfrac{(6-8)^2}{8} + ... + \dfrac{(9-8)^2}{8} \right] = 1.25$. Because 1.25 is not $\geq \chi^2_{.10,4} = 7.779$, even at level .10 H_o cannot be rejected; the data is quite consistent with the specified exponential distribution.

Chapter 13: Goodness-of-Fit Tests and Categorical Data Analysis

11.

a. The six intervals must be symmetric about 0, so denote the 4th, 5th and 6th intervals by [0, a), [a, b), [b, ∞). a must be such that $\Phi(a) = .6667\left(\frac{1}{2} + \frac{1}{6}\right)$, which from Table A.3 gives $a \approx .43$. Similarly $\Phi(b) = .8333$ implies $b \approx .97$, so the six intervals are $(-\infty, -.97)$, $[-.97, -.43)$, $[-.43, 0)$, $[0, .43)$, $[.43, .97)$, and $[.97, \infty)$.

b. The six intervals are symmetric about the mean of .5. From **a**, the fourth interval should extend from the mean to .43 standard deviations above the mean, i.e., from .5 to .5 + .43(.002), which gives [.5, .50086). Thus the third interval is [.5 − .00086, .5) = [.49914, .5). Similarly, the upper endpoint of the fifth interval is .5 + .97(.002) = .50194, and the lower endpoint of the second interval is .5 − .00194 = .49806. The resulting intervals are $(-\infty, .49806)$, $[.49806, .49914)$, $[.49914, .5)$, $[.5, .50086)$, $[.50086, .50194)$, and $[.50194, \infty)$.

c. Each expected count is $45\left(\frac{1}{6}\right) = 7.5$, and the observed counts are 13, 6, 6, 8, 7, and 5, so $\chi^2 = 5.53$. With 5 d.f., the p-value > .10, so we would fail to reject H_o at any of the usual levels of significance. There is no significant evidence to suggest that the bolt diameters are not normally distributed with $\mu = .5$ and $\sigma = .002$.

13. According to the stated model, the three cell probabilities are $(1 - p)^2$, $2p(1 - p)$, and p^2, so we wish the value of p which maximizes $(1 - p)^{2n_1}[2p(1 - p)]^{n_2} p^{2n_3}$.

Proceeding as in example 14.6 gives $\hat{p} = \dfrac{n_2 + 2n_3}{2n} = \dfrac{234}{2776} = .0843$. The estimated expected cell counts are then $n(1 - \hat{p})^2 = 1163.85$, $n[2\hat{p}(1 - \hat{p})]^2 = 214.29$,

$n\hat{p}^2 = 9.86$. This gives

$$\chi^2 = \left[\frac{(1212 - 1163.85)^2}{1163.85} + \frac{(118 - 214.29)^2}{214.29} + \frac{(58 - 9.86)^2}{9.86}\right] = 280.3. \text{ H}_o \text{ will be}$$

rejected if $\chi^2 \geq \chi^2_{\alpha,2}$, and since $\chi^2_{.01,2} = 9.210$, H_o is soundly rejected; the stated model is strongly contradicted by the data.

Chapter 13: Goodness-of-Fit Tests and Categorical Data Analysis

15. The part of the likelihood involving θ is $\left[(1-\theta)^4\right]^{n_1} \cdot \left[\theta(1-\theta)^3\right]^{n_2} \cdot \left[\theta^2(1-\theta)^2\right]^{n_3} \cdot$

$\left[\theta^3(1-\theta)\right]^{n_4} \cdot \left[\theta^4\right]^{n_5} = \theta^{n_2+2n_3+3n_4+4n_5}(1-\theta)^{4n_1+3n_2+2n_3+n_4} = \theta^{233}(1-\theta)^{367}$, so

$\ln(likelihood) = 233\ln\theta + 367\ln(1-\theta)$. Differentiating and equating to 0 yields

$\hat{\theta} = \dfrac{233}{600} = .3883$, and $(1-\hat{\theta}) = .6117$ [note that the exponent on θ is simply the total

# of successes (defectives here) in the $n = 4(150) = 600$ trials.] Substituting this $\hat{\theta}$
into the formula for p_i yields estimated cell probabilities .1400, .3555, .3385, .1433,
and .0227. Multiplication by 150 yields the estimated expected cell counts are 21.00,
53.33, 50.78, 21.50, and 3.41. the last estimated expected cell count is less than 5, so
we combine the last two categories into a single one (≥ 3 defectives), yielding
estimated counts 21.00, 53.33, 50.78, 24.91, observed counts 26, 51, 47, 26, and
$\chi^2 = 1.62$. With d.f. $= 4 - 1 - 1 = 2$, since $1.62 < \chi^2_{.10,2} = 4.605$, the p-value $> .10$,
and we do not reject H_0. The data suggests that the stated binomial distribution is
plausible.

17. $\hat{\lambda} = \dfrac{380}{120} = 3.167$, so $\hat{p} = e^{-3.167}\dfrac{(3.167)^x}{x!}$.

x	0	1	2	3	4	5	6	≥ 7
$\hat{p}$	.0421	.1334	.2113	.2230	.1766	.1119	.0590	.0427
$n\hat{p}$	5.05	16.00	25.36	26.76	21.19	13.43	7.08	5.12
obs	24	16	16	18	15	9	6	16

The resulting value of $\chi^2 = 103.98$, and when compared to $\chi^2_{.01,7} = 18.474$, it is
obvious that the Poisson model fits very poorly.

19. With $A = 2n_1 + n_4 + n_5$, $B = 2n_2 + n_4 + n_6$, and $C = 2n_3 + n_5 + n_6$, the likelihood is
proportional to $\theta_1^A\theta_2^B(1-\theta_1-\theta_2)^C$, where $A + B + C = 2n$. Taking the natural log

and equating both $\dfrac{\partial}{\partial\theta_1}$ and $\dfrac{\partial}{\partial\theta_2}$ to zero gives $\dfrac{A}{\theta_1} = \dfrac{C}{1-\theta_1-\theta_2}$ and

$\dfrac{B}{\theta_2} = \dfrac{C}{1-\theta_1-\theta_2}$, whence $\theta_2 = \dfrac{B\theta_1}{A}$. Substituting this into the first equation gives

$\theta_1 = \dfrac{A}{A+B+C}$, and then $\theta_2 = \dfrac{B}{A+B+C}$. Thus $\hat{\theta}_1 = \dfrac{2n_1+n_4+n_5}{2n}$,

$\hat{\theta}_2 = \dfrac{2n_2+n_4+n_6}{2n}$, and $(1-\hat{\theta}_1-\hat{\theta}_2) = \dfrac{2n_3+n_5+n_6}{2n}$. Substituting the observed

n_i's yields $\hat{\theta}_1 = \dfrac{2(49)+20+53}{400} = .4275$, $\hat{\theta}_2 = \dfrac{110}{400} = .2750$, and $\left(1-\hat{\theta}_1-\hat{\theta}_2\right) = .2975$,

from which $\hat{p}_1 = (.4275)^2 = .183$, $\hat{p}_2 = .076$, $\hat{p}_3 = .089$, $\hat{p}_4 = 2(.4275)(.275) = .235$,

$\hat{p}_5 = .254$, $\hat{p}_6 = .164$.

Category	1	2	3	4	5	6
np	36.6	15.2	17.8	47.0	50.8	32.8
observed	49	26	14	20	53	38

This gives $\chi^2 = 29.1$. With $\chi^2_{.01,6-1-2} = \chi^2_{.01,3} = 11.344$, and

$\chi^2_{.01,6-1} = \chi^2_{.01,5} = 15.085$, according to (14.15) H_o must be rejected since $29.1 \geq 15.085$.

21. The Ryan-Joiner test p-value is larger than .10, so we conclude that the null hypothesis of normality cannot be rejected. This data could reasonably have come from a normal population. This means that it would be legitimate to use a one-sample t test to test hypotheses about the true average ratio.

23. We want to test homogeneity of win-loss percentages across years. Minitab output appears below.

```
              C1       C2    Total
    1         73       71     144
           68.36    75.64
           0.315    0.285

    2         49       64     113
           53.64    59.36
           0.402    0.363

 Total       122      135     257

    Chi-Sq = 1.365, DF = 1, P-Value = 0.243
```

Hence, we fail to reject the null hypothesis of equal win probabilities in the (conceptual) populations of all Cubs games in 1994 and 1995. From 10.56, the test statistic was $z = -1.17$; notice that $(-1.17)^2 = 1.37$, the same as our χ^2 statistic. Also, the P-value from 10.56 was .121, or half of the P-value here. That's because 10.56 was a one-sided hypothesis test, whereas a χ^2 test is necessarily two-sided.

Chapter 13: Goodness-of-Fit Tests and Categorical Data Analysis

25. Let p_{i1} = the probability that a fruit given treatment i matures and p_{i2} = the probability that a fruit given treatment i aborts. Then H_o: $p_{i1} = p_{i2}$ for i = 1, 2, 3, 4, 5 will be rejected if $\chi^2 \geq \chi^2_{.01,4} = 13.277$.

Observed			Estimated Expected		
Matured	Aborted		Matured	Aborted	n_i
141	206		110.7	236.3	347
28	69		30.9	66.1	97
25	73		31.3	66.7	98
24	78		32.5	69.5	102
20	82		32.5	69.5	102
			238	508	746

Thus $\chi^2 = \dfrac{(141-110.7)^2}{110.7} + \ldots + \dfrac{(82-69.5)^2}{69.5} = 24.82$, which is ≥ 13.277, so H_o is rejected at level .01.

27. With p_{ij} denoting the probability of a type j response when treatment i is applied, H_o: $p_{1j} = p_{2j} = p_{3j} = p_{4j}$ for j = 1, 2, 3, 4 will be rejected at level .005 if $\chi^2 \geq \chi^2_{.005,9} = 23.587$.

$\hat{E}_{ij}$	1	2	3	4
1	24.1	10.0	21.6	40.4
2	25.8	10.7	23.1	43.3
3	26.1	10.8	23.4	43.8
4	30.1	12.5	27.0	50.5

$\chi^2 = 27.66 \geq 23.587$, so reject H_o at level .005.

29. $\chi^2 = \dfrac{(479-494.4)^2}{494.4} + \dfrac{(173-151.5)^2}{151.5} + \dfrac{(119-125.2)^2}{125.2} + \dfrac{(214-177.0)^2}{177.0} + \dfrac{(47-54.2)^2}{54.2}$

$= \dfrac{(15-44.8)^2}{44.8} + \dfrac{(172-193.6)^2}{193.6} + \dfrac{(45-59.3)^2}{59.3} + \dfrac{(85-49.0)^2}{49.0} = 64.65 \geq \chi^2_{.01,4} = 13.277$ so the independence hypothesis is rejected in favor of the conclusion that political views and level of marijuana usage are related.

240

Chapter 13: Goodness-of-Fit Tests and Categorical Data Analysis

31. Under the null hypothesis, we compute estimated cell counts by

$$\hat{e}_{ijk} = n\hat{p}_{ijk} = n\hat{p}_{i..}\hat{p}_{.j.}\hat{p}_{..k} = n\frac{n_{i..}}{n}\frac{n_{.j.}}{n}\frac{n_{..k}}{n} = \frac{n_{i..}n_{.j.}n_{..k}}{n^2}.$$ This is a 3x3x3 situation, so

there are 27 cells. Only the total sample size, n, is fixed in advance of the experiment, so there are 26 freely determined cell counts. We must estimate $p_{..1}$, $p_{..2}$, $p_{..3}$, $p_{.1.}$, $p_{.2.}$, $p_{.3.}$, $p_{1..}$, $p_{2..}$, and $p_{3..}$ but $\Sigma p_{i..} = \Sigma p_{.j.} = \Sigma p_{..k} = 1$, so only 6 independent parameters are estimated. The rule for d.f. now gives χ^2 d.f. = 26 − 6 = 20. In general, the degrees of freedom for independence in an IxJxK array equals (IJK − 1) − [(I − 1) + (J − 1) + (K − 1)] = IJK − (I+J+K) + 2.

33.

a.

Observed					Estimated Expected		
13	19	28	60		12	18	30
7	11	22	40		8	12	20
20	30	50	100				

$$\chi^2 = \frac{(13-12)^2}{12} + ... + \frac{(22-20)^2}{20} = .6806.$$ Because $.6806 < \chi^2_{.10,2} = 4.605$, H_o is not rejected.

b. Each observation count here is 10 times what it was in **a**, and the same is true of the estimated expected counts, so now $\chi^2 = 6.806 \geq 4.605$, and H_o is rejected. With the much larger sample size, the departure from what is expected under H_o, the independence hypothesis, is statistically significant – it cannot be explained just by random variation.

c. The observed counts are $.13n, .19n, .28n, .07n, .11n, .22n$, whereas the estimated expected $\frac{(.60n)(.20n)}{n} = .12n, .18n, .30n, .08n, .12n, .20n$, yielding

$\chi^2 = .006806n$. H_o will be rejected at level .10 iff $.006806n \geq 4.605$, i.e., iff $n \geq 676.6$, so the minimum n = 677.

Chapter 13: Goodness-of-Fit Tests and Categorical Data Analysis

35.

   a. Minitab output for the test of gender homogeneity among ranks appears below. We see the test statistic is $\chi^2 = 6.454$, with a P-value of 0.04 at 2 df. Thus, at the $\alpha = .05$ level, we reject the null hypothesis and conclude that the proportion of Professors, Associate Professors, and Assistant Professors who are female are different.

```
           M       F   Total
  1       25       9      34
        21.42   12.58
        0.598   1.019

  2       20       8      28
        17.64   10.36
        0.316   0.538

  3       18      20      38
        23.94   14.06
        1.474   2.510

Total     63      37     100

Chi-Sq = 6.454, DF = 2, P-Value = 0.040
```

   b. Minitab output for the appropriate logistic regression (with Success = Female, 1 = Professor, 2 = Assoc. Prof., 3 = Asst. Prof.) is shown below. From the output, the test of H_0: $\beta_1 = 0$ yields $z = 2.29$ with a P-value 0.022. Thus, again we reject the hypothesis that rank and gender are unrelated. In particular, the output shows that the odds a faculty member is female increases significantly as we continue <u>down</u> the table (i.e., down the ranks from Professor to Assistant).

```
Logistic Regression Table

                                              Odds      95% CI
Predictor      Coef    SE Coef     Z      P  Ratio   Lower  Upper
Constant   -1.76732   0.593645  -2.98  0.003
C8          0.589227  0.257333   2.29  0.022   1.80    1.09   2.98
```

   c. Yes: with the extra assumption of order among the factor, we anticipate the P-value from logistic regression will be lower than the P-value from the chi-square test.

   d. The gender imbalance among faculty is less pronounced in lower ranks and more pronounced in higher ranks. This suggests more female faculty are being hired than before. If all these faculty remain at the university, in 10-15 years time we will see noticeably less gender imbalance among the full professors.

Chapter 13: Goodness-of-Fit Tests and Categorical Data Analysis

37. Let p_{i1} = the proportion of fish receiving treatment i (i = 1, 2, 3) who are parasitized. We wish to test H_0: $p_{1j} = p_{2j} = p_{3j}$ for j = 1, 2. With df = (2 – 1)(3 – 1) = 2, H_0 will be rejected at level .01 if $\chi^2 \geq \chi^2_{.01,2} = 9.210$.

Observed			Estimated Expected	
30	3	33	22.99	10.01
16	8	24	16.72	7.28
16	16	32	22.29	9.71
62	27	89		

This gives $\chi^2 = 13.1$. Because $13.1 \geq 9.210$, H_0 should be rejected. The proportion of fish that are parasitized does appear to depend on which treatment is used.

39.

    a. H_0: The proportion of Late Game Leader Wins is the same for all four sports; H_a: The proportion of Late Game Leader Wins is not the same for all four sports. With 3 df, the computed $\chi^2 = 10.518$, and the p-value < .015 < .05, we would reject H_0. There appears to be a relationship between the late-game leader winning and the sport played.

    b. Quite possibly: Baseball had many fewer late-game leader losses than expected.

41. The estimated expected counts are displayed below, from which $\chi^2 = 197.70$. A glance at the 6 df row of Table A.7 shows that this test statistic value is highly significant – the hypothesis of independence is clearly implausible.

	Estimated Expected			
	Home	Acute	Chronic	
15 – 54	90.2	372.5	72.3	535
55 – 64	113.6	469.3	91.1	674
65 – 74	142.7	589.0	114.3	846
> 74	157.5	650.3	126.2	934
	504	2081	404	2989

43. The accompanying table contains both observed and estimated expected counts, the latter in parentheses.

	Age					
Want	127	118	77	61	41	424
	(131.1)	(123.3)	(71.7)	(55.1)	(42.8)	
Don't	23	23	5	2	8	61
	(18.9)	(17.7)	(10.3)	(7.9)	(6.2)	
	150	141	82	63	49	485

Chapter 13: Goodness-of-Fit Tests and Categorical Data Analysis

This gives $\chi^2 = 11.60 \geq \chi^2_{.05,4} = 9.488$. At level .05, the null hypothesis of independence is rejected, though it would not be rejected at the level .01 (.01 < p-value < .025).

45.

a.

obsv	22	10	5	11
exp	13.189	10	7.406	17.405

H_0: probabilities are as specified.
H_a: probabilities are not as specified.

Test Statistic: $\chi^2 = \dfrac{(22-13.189)^2}{13.189} + \dfrac{(10-10)^2}{10} + \dfrac{(5-7.406)^2}{7.406} + \dfrac{(11-17.405)^2}{17.405}$

$= 5.886 + 0 + 0.782 + 2.357 = 9.025$. Rejection Region: $\chi^2 > \chi^2_{.05,3} = 7.815$.

Since $9.025 > 7.815$, we reject H_0. The model postulated in the exercise is not a good fit.

b.

p_i	0.45883	0.18813	0.11032	0.24272
exp	22.024	9.03	5.295	11.651

$\chi^2 = \dfrac{(22-22.024)^2}{22.024} + \dfrac{(10-9.03)^2}{9.03} + \dfrac{(5-5.295)^2}{5.295} + \dfrac{(11-11.651)^2}{11.651}$

$= .0000262 + .1041971 + .0164353 + .0363746 = .1570332$

With the same rejection region as in a, we do not reject the null hypothesis. This model does provide a good fit.

47.

a. H_0: $p_0 = p_1 = \ldots = p_9 = .10$ vs H_a: at least one $p_i \neq .10$, with df = 9.

b. H_0: $p_{ij} = .01$ for i and $j = 0,1,2,\ldots,9$ vs H_a: at least one $p_{ij} \neq .01$, with df = 99.

c. For this test, the number of p's in the Hypothesis would be $10^5 = 100{,}000$ (the number of possible combinations of 5 digits). Using only the first 100,000 digits in the expansion, the number of non-overlapping groups of 5 is only 20,000. We need a much larger sample size!

d. Based on these p-values, we could conclude that the digits of π behave as though they were randomly generated.

Chapter 14: Alternative Approaches to Inference

1. We test $H_0 : \mu = 100$ vs. $H_a : \mu \neq 100$. The test statistic is $s_+ =$ sum of the ranks associated with the positive values of $(x_i - 100)$, and we reject H_o at significance level .05 if $s_+ \geq 64$. (from Table A.13, n = 12, with $\alpha/2 = .026$, which is close to the desired value of .025), or if $s_+ \leq \dfrac{12(13)}{2} - 64 = 78 - 64 = 14$.

x_i	$(x_i - 100)$	ranks
105.6	5.6	7*
90.9	-9.1	12
91.2	-8.8	11
96.9	-3.1	3
96.5	-3.5	5
91.3	-8.7	10
100.1	0.1	1*
105	5	6*
99.6	-0.4	2
107.7	7.7	9*
103.3	3.3	4*
92.4	-7.6	8

$S_+ = 27$, and since 27 is neither ≥ 64 nor ≤ 14, we do not reject H_o. There is not enough evidence to suggest that the mean is something other than 100.

3. We test $H_0 : \mu = 7.39$ vs. $H_a : \mu \neq 7.39$, so a two tailed test is appropriate. With n = 14 and $\alpha/2 = .025$, Table A.13 indicates that H_o should be rejected if either $s_+ \geq 84 or \leq 21$. The $(x_i - 7.39)$'s are -.37, -.04, -.05, -.22, -.11, .38, -.30, -.17, .06, -.44, .01, -.29, -.07, and -.25, from which the ranks of the three positive differences are 1, 4, and 13. Since $s_+ = 18 \leq 21$, H_o is rejected at level .05

5. The data is paired, and we wish to test $H_0 : \mu_D = 0$ vs. $H_a : \mu_D \neq 0$. With n = 12 and $\alpha = .05$, H_o should be rejected if either $s_+ \geq 64 or$ if $s_+ \leq 14$.

d_i	-.3	2.8	3.9	.6	1.2	-1.1	2.9	1.8	.5	2.3	.9	2.5
rank	1	10*	12*	3*	6*	5	11*	7*	2*	8*	4*	9*

$s_+ = 72$, and $72 \geq 64$, so H_o is rejected at level .05. In fact for $\alpha = .01$, the critical value is c = 71, so even at level .01 H_o would be rejected.

Chapter 14: Alternative Approaches to Inference

7. We wish to test $H_0 : \mu = 75$ vs. $H_a : \mu > 75$. Since n = 25 the large sample approximation is used, so H_0 will be rejected at level .05 if $z \geq 1.645$. The $(x_i - 75)'s$ are −5.5, -3.1, -2.4, -1.9, -1.7, 1.5, -.9, -.8, .3, .5, .7, .8, 1.1, 1.2, 1.2, 1.9, 2.0, 2.9, 3.1, 4.6, 4.7, 5.1, 7.2, 8.7, and 18.7. The ranks of the positive differences are 1, 2, 3, 4.5, 7, 8.5, 8.5, 12.5, 14, 16, 17.5, 19, 20, 21, 23, 24, and 25, so s_+ = 226.5 and $\frac{n(n+1)}{4} = 162.5$. Expression (14.2) for σ^2 should be used (because of the ties):

$\tau_1 = \tau_2 = \tau_3 = \tau_4 = 2$, so $\sigma_{s_+}^2 = \dfrac{25(26)(51)}{24} - \dfrac{4(1)(2)(3)}{48} = 1381.25 - .50 = 1380.75$

and $\sigma = 37.16$. Thus $z = \dfrac{226.5 - 162.5}{37.16} = 1.72$. Since $1.72 \geq 1.645$, H_0 is rejected.

$p - value \approx 1 - \Phi(1.72) = .0427$. The data indicates that true average toughness of the steel does exceed 75.

9. The ordered combined sample is 163(y), 179(y), 213(y), 225(y), 229(x), 245(x), 247(y), 250(x), 286(x), and 299(x), so w = 5 + 6 + 8 + 9 + 10 = 38. With m = n = 5, Table A.14 gives the upper tail critical value for a level .05 test as 36 (reject H_0 if W ≥ 36). Since 38 ≥ 36, H_0 is rejected in favor of H_a.

11. The hypotheses of interest are $H_0 : \mu_1 - \mu_2 = 1$ vs. $H_a : \mu_1 - \mu_2 > 1$, where 1(X) refers to the original process and 2 (Y) to the new process. Thus 1 must be subtracted from each x_1 before pooling and ranking. At level .05, H_0 should be rejected in favor of H_a if w ≥ 84.

x − 1	3.5	4.1	4.4	4.7	5.3	5.6	7.5	7.6
rank	1	4	5	6	8	10	15	16
y	3.8	4.0	4.9	5.5	5.7	5.8	6.0	7.0
rank	2	3	7	9	11	12	13	14

Since w = 65, H_0 is not rejected.

13.

a.

X	rank	Y	rank
0.43	2	1.47	9
1.17	8	0.8	7
0.37	1	1.58	11
0.47	3	1.53	10
0.68	6	4.33	16
0.58	5	4.23	15
0.5	4	3.25	14
2.75	12	3.22	13

We verify that w = sum of the ranks of the x's = 41.

b. We are testing $H_0 : \mu_1 - \mu_2 = 0$ vs. $H_a : \mu_1 - \mu_2 < 0$. The reported p-value (significance) is .0027, which is < .01 so we reject H_0. There is evidence that the distribution of good visibility response time is to the left (or lower than) that response time with poor visibility.

15. Here m = n = 10 > 8, so we use the large-sample test statistic from pp. 680-681. $H_0 : \mu_1 - \mu_2 = 0$ will be rejected at level .01 in favor of $H_a : \mu_1 - \mu_2 \neq 0$ if either $z \geq 2.58$ or $z \leq -2.58$. Identifying X with orange juice, the X ranks are 7, 8, 9, 10, 11, 16, 17, 18, 19, and 20, so w = 135. With $\dfrac{m(m+n+1)}{2} = 105$ and

$$\sqrt{\frac{mn(m+n+1)}{12}} = \sqrt{175} = 13.22, \quad z = \frac{135-105}{13.22} = 2.27.$$ Because 2.27 is neither ≥ 2.58 nor ≤ -2.58, H_0 is not rejected. $p-value \approx 2(1 - \Phi(2.27)) = .0232$.

17. n = 8, so from Table A.15, a 95% C.I. (actually 94.5%) has the form $\left(\bar{x}_{(36-32+1)}, \bar{x}_{(32)} \right) = \left(\bar{x}_{(5)}, \bar{x}_{(32)} \right)$. It is easily verified that the 5 smallest pairwise averages are $\dfrac{5.0+5.0}{2} = 5.00$, $\dfrac{5.0+11.8}{2} = 8.40$, $\dfrac{5.0+12.2}{2} = 8.60$, $\dfrac{5.0+17.0}{2} = 11.00$, and $\dfrac{5.0+17.3}{2} = 11.15$ (the smallest average not involving 5.0 is $\bar{x}_{(6)} = \dfrac{11.8+11.8}{2} = 11.8$), and the 5 largest averages are 30.6, 26.0, 24.7, 23.95, and 23.80, so the confidence interval is (11.15, 23.80).

Chapter 14: Alternative Approaches to Inference

19. With $n = 8$, Table A.15 gives $c = 32$, and so a 94.5% confidence interval for μ_D is $(\bar{x}_{(5)}, \bar{x}_{(32)})$. The five smallest pairwise averages are -.71, -.68, -.65, -.615, -.585; the five largest pairwise averages are .2, .195, .19, .03, .025. The desired C.I. is thus $(-.585, .025)$.

21. $m = n = 5$ and from Table A.16, $c = 21$ and the 90% (actually 90.5%) interval is $(d_{ij(5)}, d_{ij(21)})$. The five smallest $x_i - y_j$ differences are $-18, -2, 3, 4, 16$ while the five largest differences are $136, 123, 120, 107, 87$ (construct a table), so the desired interval is $(16, 87)$.

23. Let $y_1, \ldots, y_n$ be our random sample of Bernoulli observations, so that $f(y \mid p) = \prod p^{y_i}(1-p)^{1-y_i} = p^{\Sigma y_i}(1-p)^{n-\Sigma y_i} = p^x(1-p)^{n-x}$ with $x = \sum y_i$. Dropping constants for simplicity's sake, Bayes' rule says $h(p \mid y) = \dfrac{f(y \mid p)g(p)}{\int f(y \mid p)g(p)dp}$

$\overset{p}{\propto} f(y \mid p)g(p) \overset{p}{\propto} p^x(1-p)^{n-x} \, p^{a-1}(1-p)^{b-1} = p^{x+a-1}(1-p)^{n-x+b-1}$, the kernel of a Beta pdf with parameters $x+a$ and $n-x+b$. This is, not surprisingly, identical to the result of Example 14.7.

25.

    a. From Example 14.7, the posterior distribution of p is Beta with parameters $a+x = 1+n$ and $b+n-x = 1+n-n = 1$. Hence, the posterior mean of p is $(1+n)/(1+n+1) = (n+1)/(n+2)$.

    b. Imagine two prior trials, one success and one failure. Then we observe n successes in the next n trials, for $n+1$ total successes out of these combined $n+2$ trials. The relative frequency of successes is then $(n+1)/(n+2)$.

    c. This seems like an odd, and arbitrary, application of Laplace's idea. Why start with two days, with the sun rising on only one of them? Then, no matter how many days follow with the sun rising, we still include those two days on which the sun rose only once.

27. Write $r = \beta/\alpha$, so that $Y = rX/(1-X)$. Solving for X gives $x = y/(r+y)$ and $dx/dy = r/(r+y)^2$. Suppressing constants for notational ease, the pdf of Y is $f_X(y/(r+y))|dx/dy| \propto$

$\left(\dfrac{y}{r+y}\right)^{\alpha-1}\left(1-\dfrac{y}{r+y}\right)^{\beta-1}\dfrac{r}{(r+y)^2} \propto \dfrac{y^{\alpha-1}}{(r+y)^{\alpha+\beta}} \propto \dfrac{y^{v_1-1}}{(v_2+v_1 y)^{v_1+v_2}}$ under the

substitutions $\alpha = v_1/2$ and $\beta = v_2/2$. This is the kernel of the F distribution with parameters v_1 and v_2 (i.e., the part of the pdf without the constants). Therefore, $Y \sim F(v_1, v_2)$.

Chapter 14: Alternative Approaches to Inference

29. In this setting, $v_1 = 2\alpha = 80$ and $v_2 = 2\beta = 60$.

a. Table A.9 does not provide the F(80,60) distribution, but we can estimate: $F_{.05,80,60} \approx 1.5$, and $F_{.95,80,60} = 1/F_{.05,60,80} \approx 1/1.49 = .67$. From Exercise 27, this suggests a 95% credibility interval for p of the form $.67 \le (p/40)/((1-p)/30) \le 1.5$. Solving for p yields $\dfrac{4(.67)}{3+4(.67)} \le p \le \dfrac{4(1.5)}{3+4(1.5)}$, or $.47 \le p \le .67$.

b. From Minitab, the .05 and .95 quantiles of the Beta(40,30) distribution are .4736 and .6669, respectively. These are a match to the values in (a).

31. $f(x \mid \lambda) = \displaystyle\prod \frac{e^{-\lambda}\lambda^{x_i}}{x_i!} = \frac{e^{-n\lambda}\lambda^{y}}{\prod x_i!}$ with $y = \sum x_i$. Dropping constants for simplicity's

sake, Bayes' rule says $h(\lambda \mid x) \overset{\lambda}{\propto} f(x \mid \lambda)g(\lambda) \overset{\lambda}{\propto} e^{-n\lambda}\lambda^{y}\,\lambda^{\alpha-1}e^{-\lambda/\beta} =$
$\lambda^{y+\alpha-1}e^{-\lambda(n+1/\beta)}$, the kernel of a Gamma pdf with parameters y+α and $1/(n + 1/\beta)$.

33.

a. To start, $A = .05/(1-.05) = 1/19$ and $B = (1-.05)/.05 = 19$. The likelihood ratio

is $\lambda_n = \dfrac{\exp(-\frac{1}{2(15)^2}\sum(x_i - 110)^2)}{\exp(-\frac{1}{2(15)^2}\sum(x_i - 100)^2)} = \exp(\frac{1}{450}[20\sum x_i - 2100n])$ after expanding

the quadratic terms in the numerator and denominator. Thus, we use Wald's procedure until we cross one of two boundaries: Reject H$_0$ if $\lambda_n \ge B = 19$, which is equivalent to $\sum x_i \ge 105n + 66.25$; Accept H$_0$ if $\lambda_n \le A = 1/19$, which is equivalent to $\sum x_i \le 105n - 66.25$.

b. We stop at $n = 7$: $\sum x_i = 812$, and the upper stopping rule is 105(7)+66.25 = 801.25.

c. From (a), the relevant log-likelihood is $\ln[\exp(\frac{1}{450}[20X_i - 2100])] =$
$\frac{1}{450}[20X_i - 2100]$, whose expected value is $\frac{1}{450}[20\mu - 2100]$. When H$_0$ is true,

$E(N) = \dfrac{.05\ln(19) + .95\ln[1/19]}{\frac{1}{450}[20(100) - 2100]} = 11.925$. When H$_a$ is true, E(N) =

$\dfrac{.95\ln(19) + .05\ln[1/19]}{\frac{1}{450}[20(110) - 2100]} = 11.925$. The normal distribution is symmetric

regardless of μ, and with the same α and β specified, these expectations will be the same.

d. The required fixed sample size is $\left(\dfrac{15(1.645+1.645)}{100-110}\right)^2 = 24.35$. This is twice as high as the value from the SPRT test in (c).

Chapter 14: Alternative Approaches to Inference

35.

    a. Substitute $y = 2k - n$, or $k = (y+n)/2$. If the bounds in (14.11) are abbreviated $lb < k < ub$, the bounds for continued sampling in terms of y are then $2lb - n < y < 2ub - n$.

    b. The bounds from Example 14.9 become $2(.631n - 2.704) - n < y < 2(.631n + 3.311) - n$, or $.262n - 5.408 < y < .262n + 6.622$.

    c. The graph below includes y (the jagged line), and the lines $.262n - 5.408$ and $.262n + 6.622$. For each success, y (the jagged line) increases by 1; for each failure, y decreases by 1. Notice that the graph of y violates the boundaries at $n = 23$, just as in Example 14.9.

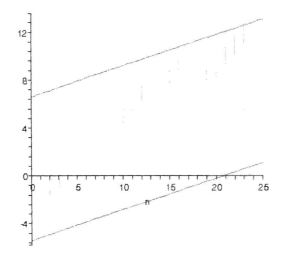

37. The Wilcoxon signed-rank test will be used to test $H_0 : \mu_D = 0$ vs. $H_0 : \mu_D \neq 0$, where μ_D = the difference between expected rate for a potato diet and a rice diet. From Table A.11 with $n = 8$, H_o will be rejected if either $s_+ \geq 32$ or $s_+ \leq \dfrac{8(9)}{2} - 32 = 4$. The $d_i's$ are (in order of magnitude) .16, .18, .25, -.56, .60, .96, 1.01, and −1.24, so $s_+ = 1 + 2 + 3 + 5 + 6 + 7 = 24$. Because 24 is not in the rejection region, H_o is not rejected.

Chapter 14: Alternative Approaches to Inference

39.

    a.   With "success" as defined, then Y is a binomial with n = 20. To determine the binomial proportion "p" we realize that since 25 is the hypothesized median, 50% of the distribution should be above 25, thus p = .50. From the Binomial Tables (Table A.1) with n = 20 and p = .50, we see that
$$\alpha = P(Y \geq 15) = 1 - P(Y \leq 14) = 1 - .979 = .021.$$

    b.   From the same binomial table as in **a**, we find that
$$P(Y \geq 14) = 1 - P(Y \leq 13) = 1 - .942 = .058 \text{ (as close as we can get to .05), so c}$$
= 14. For this data, we would reject H_o at level .058 if $Y \geq 14$. Y = (the number of observations in the sample that exceed 25) = 12, and since 12 is not ≥ 14, we fail to reject H_o.

41.   The table below gives the ranked form of the data.

1:	1	2	3	4	5	10	24	r = 49	$\bar{r} = 7.00$
2:	6	8	9	13	17	21	22	r = 96	$\bar{r} = 13.71$
3:	11	12	15	16	18	20	25	r = 117	$\bar{r} = 16.71$
4:	7	14	19	26	29	32	33	r = 160	$\bar{r} = 22.86$
5:	23	27	28	30	31	34	35	r = 208	$\bar{r} = 29.71$

With J = 7 for each group and N = 35, $k = \dfrac{12}{35(36)} \sum 7(\bar{r}_i - 18)^2 = 20.12$. At the $\alpha =$ .01 level, $\chi^2_{.01,4} = 13.277$, and since $20.12 \geq 13.277$, H_0 is rejected and we conclude that expected axial stiffness does depend on plate length.

43.   The table below gives the data ranked within each block.

Fear	4	4	3	4	1	4	4	3	$\bar{r} = 27/8$
Happ.	3	2	2	1	4	3	1	4	$\bar{r} = 20/8$
Depr.	1	3	4	2	3	2	2	2	$\bar{r} = 19/8$
Calm.	2	1	1	3	2	1	3	1	$\bar{r} = 14/8$

With J = 8 and I = 4, $Fr = \dfrac{12(8)}{4(5)} \sum (\bar{r}_i - 2.5)^2 = 6.45$. At the $\alpha = .05$ level, $\chi^2_{.05,3} =$ 7.815, and since 6.45 < 7.815, H_0 is <u>not</u> rejected. There is no statistically significant evidence that average skin potential depends on which emotion is requested.

Chapter 14: Alternative Approaches to Inference

45.

Sample:	y	x	y	y	x	x	x	y	y
Observations:	3.7	4.0	4.1	4.3	4.4	4.8	4.9	5.1	5.6
Rank:	1	3	5	7	9	8	6	4	2

The value of W' for this data is $w' = 3 + 6 + 8 + 9 = 26$. At level .05, the critical value for the upper-tailed test is (Table A.14, m = 4, n = 5) c = 27 (α = .056). Since 26 is not $\geq$ 27, H_o cannot be rejected at level .05.

Made in the USA
Lexington, KY
14 September 2011